New Generation of Electric Vehicles

New Generation of Electric Vehicles

Editor

Johanna Adison

scitus
academics

New Generation of Electric Vehicles
Edited by **Johanna Adison**

ISBN: 978-1-68117-655-0
Library of Congress Control Number: 2016946522

SCITUS Academics LLC,
616, Corporate Way, Suite 2, 4766,
Valley Cottage, NY 10989

www.scitusacademics.com

Printed in United States of America on Acid Free Paper

Preface

There is an unprecedented surge of interest in and the desire of producing gas-electric hybrid (HEVs) and all-electric vehicles (EVs) by the governments and public in the industrialized countries. This surge of interest and desire is attributed to several factors relating to the life-threatening air pollution in many parts of the world, the high petroleum price, conflicts in oil producing Middle East regions and the alarming trend of global warming due to rapid increase of greenhouse gas emissions by all sources, including those from transportation. Full electric vehicles use no gasoline; they are powered by a high-voltage electric motor and battery pack. All the elements are now in place to tackle fairly complex static rigid body problems. Battery electric vehicles do not have an internal combustion engine and use no on-board gasoline. Instead, they use a high-voltage electric motor, which gets its power from a high-voltage battery pack. The primary benefit of BEVs is that they completely eliminate carbon dioxide (CO_2) and other emissions directly from the vehicle.

Therefore, it is vital to raise global ecological awareness and wider public education regarding ecology. Goal of this book is to bring closer to the readers new drive technologies that are intended to environment and nature protection.

New Generation of Electric Vehicles presents modern technique achievements and technologies applied in the implementation of electric vehicles. Special attention was paid to energy efficiency of electrical vehicles. Also today's trends, mathematical models and computer design elements of future cars are presented.

Table of Contents

Chapter 1

Energy Efficiency of Electric Vehicles

Zoran Stevic[1] and Ilija Radovanovic[2]

[1] Technical Faculty in Bor, University of Belgrade, Serbia

[2] Innovation Center of School of Electrical Engineering,
University of Belgrade, Serbia

1. INTRODUCTION

In this chapter, the most important possibilities for increasing energy efficiency of electric vehicles would be considered, regarding energy savings accumulated in the vehicle itself and increasing the range of performances of the cars with given initial resources. Some of the possibilities that should provide such a progress nowadays are:

- Using energy under braking
- Using waste heat energy
- Additional supply by solar cells
- Improved mechanical energy transmission system
- Improved cars shell design
- Increasing of efficiency of power convertors
- Special design of electric engines
- Using supercapacitors, fuel cells and new generation batteries

- Route selection on the criterion of minimum consumption in real time
- Parameter monitoring inside and outside of the vehicle and computerized system control with optimization of energy consumption

Today, the problem of energy becomes so important that an entire industry is turning towards clean, renewable energy (solar energy, wind energy, etc.). Prototypes of hybrid vehicles with the announcement of mass production scheduled for the near future have become everyday occurrence. In addition, many cars are designed to use only electricity as motive power, which reduces emissions to zero.

Photo cells in a glass roof generate electricity, even at lower intensity of solar radiation; this current operates using a fan in a vehicle. In this way the vehicle interior has a constant supply of fresh air and pleasant temperatures (up to 50% lower), although the motor vehicle is turned off so that fuel economy is evident. The solar roof is only the beginning, while the development of city cars is going toward solar vehicles prototype.

A solar vehicle is an electric vehicle powered completely or significantly by direct solar energy. Usually, photovoltaic (PV) cells contained in solar panels convert the sun's energy directly into electric energy. The term "solar vehicle" usually implies that solar energy is used to power all or part of a vehicle's propilsion. Solar power may be also used to provide power for communications or controls or other auxiliary functions.

Another concept that has been developing over the years is a kinetic energy recovery system, often known simply as KERS. KERS is an automotive system for recovering a moving vehicle's kinetic energy under braking. The recovered energy is stored in a reservoir (for example a flyeheel or a batterry or supercapacitor) for later use under acceleration. Electrical systems use a motor-generator incorporated in the car's transmission which converts mechanical energy into electrical energy and vice versa. Once the energy has been harnessed, it is stored in a battery and released when required. The mechanical KERS system utilizes flywheel technology to recover and store a moving vehicle's kinetic energy which is otherwise wasted when the vehicle is decelerated. Compared to the alternative of electrical-battery systems, the mechanical KERS system provides a significantly more compact, efficient, lighter and environmentally-friendly solution. There is one other option available - hydraulic KERS, where braking energy is used to accumulate hydraulic pressure which is then sent to the wheels when required.

Development of new components, improved connections and electric engine control algorithms allow increase of efficiency of power convertors, therefore electric engine itself, to the maximum theoretical limits. New generation improvements of electric engine system has an impact on price, however investment quickly pays off during operating.

Major efforts are invested in the development of high energy density batteries with minimum ESR. Also, current research show that fuel cells have reached needed performances for commercial use in electric vehicles. Supercapacitors that provide high power density increase the acceleration of vehicle as well as collecting all the energy from instant braking, therefore improvements of the characteristics of power supply are made.

Modern electric vehicles have full information system that has constant modifications and does monitoring of inside and outside parameters in order to achieve maximum energy savings. Except for smart sensors, it is highly important to process GPS signals and route selection on the criterion of minimum energy consumption.

By combining these technologies, concepts and their improvements, we are slowly going towards energy-efficient vehicles which will greatly simplify our lives in the future.

2. ELECTRICAL LOSSES REDUCTION IN EV

2.1. Energy Efficiency Of The Converters

Increasing of the energy efficiency of the convertors can be achieved by optimizing their configuration and control, as well as choosing the adequate component. Converter configuration depends on the type of the electric motor (DC or AC), possible recovery energy braking, drive dynamics etc.

For DC motor supply there are mostly used chopper voltage reducers, so they will be considered here.Figure 7 shows simplified presentation of the chopper supply of a DC motor. Chopper is shown as ideal breaker controlled by voltage (U_{up}), so it can control switching on (T_{ON}) and switch-off (T_{off}) exiting voltage (U_{do}). For all four quadrant operation transistor bridge as shown in fig. 1 can be used [1].

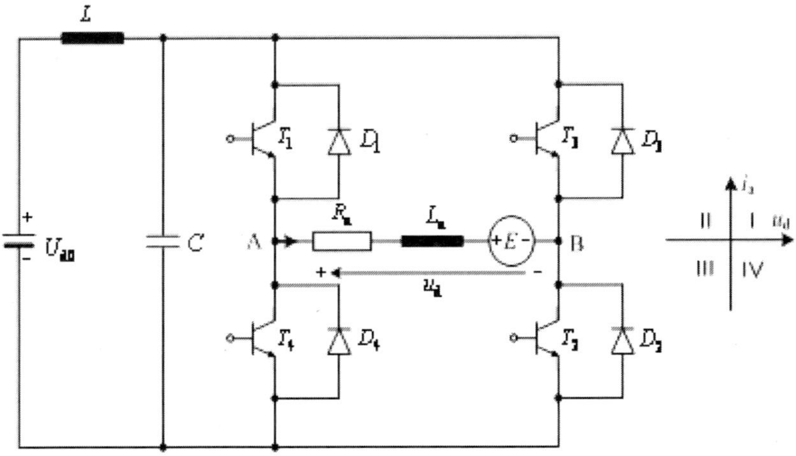

Figure 1. Transistor Bridge

By switching on transistor pairs T_1-T_2 or T_3-T_4 positive or negative polarity of motor voltage u_d is provided. To close motor current at null or reverse polarization, diodes D_1 to D_4 are provided. Converter part of the AC drive of the vehicle consists of the inverters, regulators and control set. The inverter is part of the drive inverter that inverts DC voltage to AC voltage necessary waveform to ensure the required control electric motors. Three-phase inverter consists of three inverter bridges with two switching elements in each bridge, therefore, a total of six switches. By controlling the moments of switching of the particular switches, and by controlling the length of their involvement, the appropriate waveforms at the output of inverter are achieved.

General modern circuit for speed regulation of DC motor is shown in figure 2. Reference rotary speed W_{ref} is set and also maximum armature current I_{amax} and their actual values are monitored and also brought into regulator which outputs present command values for excitation actuators and inductor [1-2].

Out of base range (for speeds above nominal) method of reduced field is used so among basic values excitation current, i_f, is monitored. Apart from classic PID action, regulating algorithm comprises other tasks (actuator command input adaptation, change of regulating method in accordance with the given speed, alarms etc.). Standard way of regulating DC drives, cascade regulation, consists of two feedbacks: internal – current and external – speed.

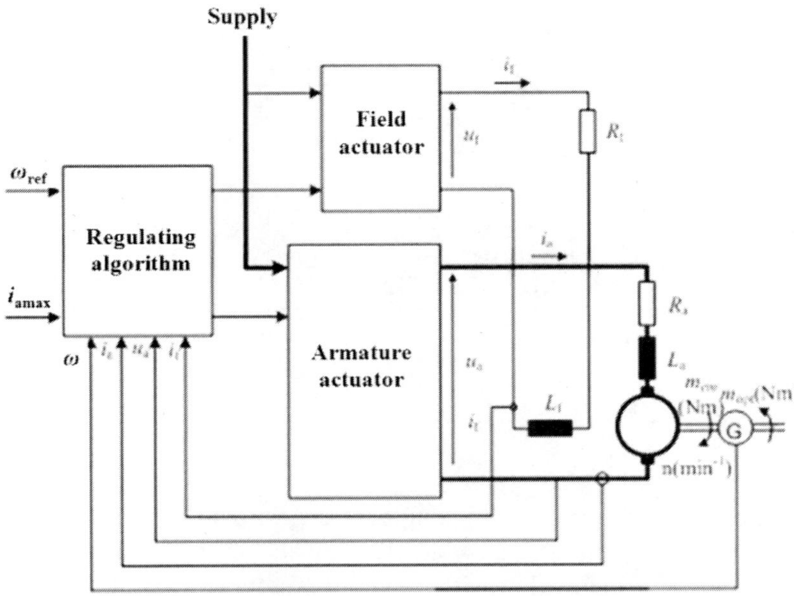

Figure 2. Circuit for speed regulation of DC motor with independent field

Asynchronous motor at constant frequency and amplitude of supply voltage rotor speed depends of load torque, which requires complicated governing algorithms in case when precise speed control and/or position. This phenomenon is a consequence of principle of asynchronous motor, and it is electromagnetic induction, which requires difference in between rotor speed and rotary magnetic field generated by stator to create electromagnetic torque. Electronics that creates algorithms mentioned was expensive earlier and such a use of asynchronous motors was difficult, but today with cheaper electronics components and use of microprocessors for regulating algorithms they are more often used.

Figure 3 represents block-diagram of regulated drive for AC motor. Depending on use and requirements, some of feedbacks and regulators can be left out. Power block (converter + motor) has two input and five output values. Input (command) parameters are effective polyphase supply voltage U_d and frequency Ws. Output (regulated) values are motor current Is, flux w, position O, rotary frequency w and torque me. Each of those has proper regulator in negative feedback, in order as shown in figure 10.

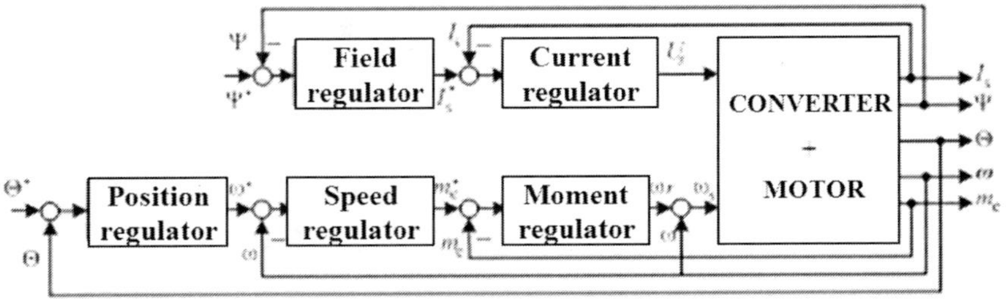

Figure 3. Block diagram of AC motor regulator

Regulation (close-loop control) comprises control with negative feedback, or feedbacks, by means of which, by means of measuring regulated parameters and comparing with required (reference) parameters those values, is acted upon command parameters, so it is automatically achieved ahead defined values of controlled values [1-2].

There may be a large energy saving by selecting the suitable power switching elements, which development is in high prosperity. As switch elements in the inverters and choppers high-power bipolar transistors, MOS (Metal Oxide Semiconductor) transistors or IGBTs (Insulated Gate Bipolar Transistor) are used. High-power bipolar transistors have very low collector-emitter resistance in the conducting state, while their control must provide sufficient supply base, it is required a relatively high power for control. On the other hand, MOS transistors have very high input resistance, and to control them it is just enough to provide the appropriate value of the voltage between the gate and source. Therefore the MOS transistor control current is almost zero and there is no power dissipation in the control circuit. Lack of MOS transistors is relatively high resistance in ON state. IGBT belongs to the family BiMOS transistors and combines these fine qualities of high-power bipolar and MOS transistors [2].

Development of multi-axis distributed control systems where sensors, actuators and controller are distributed across networks. System features system synchronized control and high speed serial communications using fiber optic channels for noise immunity. In addition, communication protocols have been developed that monitor data integrity and can sustain operation in the event of a temporary loss of communication channel. Engineers can design a system to meet exact customer requirements (fig. 4) [2].

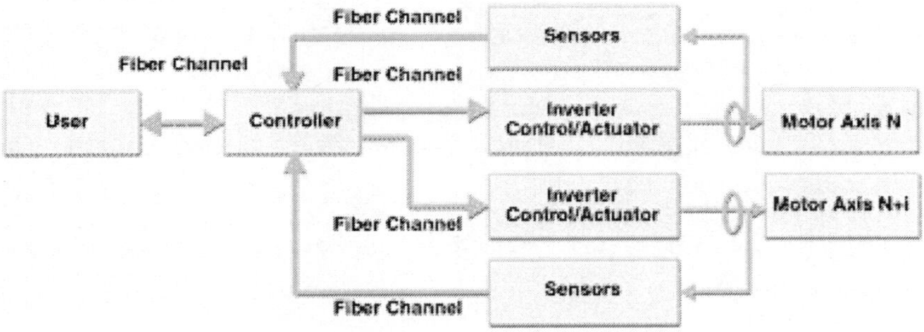

Figure 4. Distributed control

In this way, the optimization of the drive by the criteria of the dynamics and energy efficiency, while following the user's request. For supply of certain components, particularly in hybrid vehicles, high power supplies of constant current or current impulses are needed. Precise management and optimization of such sources today is exclusively microprocessor controlled [3].

2.2. Energy Efficiency Of The Electric Motor

The electric motor is the most important part of the electrical drive and the last link in the chain of energy conversion. DC motors because of their good qualities, control of the rotation speed and control of the torque, for a long time have been irreplaceable part of the controlled electric motor drives. In recent years, thanks to the advanced control techniques, asynchronous motors take place of the DC motors in regulated drives because of its good properties (robustness, lower maintenance requirements and their appliances in explosive environments, which are especially important in the case of hybrid vehicles).

Electric motor drive is designed and optimized starting from the known parameters of the engine. The latest methods for minimizing the power losses in real-time by reducing the level of flux does not require knowing of all engine parameters, and can be applied to asynchronous motor drives with scalar and vector control. Optimization of efficiency of asynchronous motors is based on adaptive adjustment of flux levels in order to determine the optimum operating point by minimizing losses [2].

Losses due to higher harmonics have to be taken into account when determining the degree of efficiency of the entire drive. The voltage at the output of the inverter is considered ideal sinusoidal in the case of control structure developing and the produced effects of higher harmonics are subsequently taken into account.

Any well-designed controller for optimization should meet the following requirements [2]:

- to determine the optimal operating point for each speed and each load torque of the defined areas of work;
- the duration of the optimization process is as short as possible;
- to have a minimum number of sensors required;
- to be easy to use;
- that it can be applied to any standard electric motor drive;
- that it can be applied to any type of engine if the only known data are on motor nameplate;
- to demonstrates a high degree of robustness in the case of disruption load torque;
- Demonstrates a high degree of robustness in the case of motor parameter variations.

Beside the standard electric motors, solutions specially made for EV are developing. Therefore, switched reluctance motor (SRM) is gaining much interest as a candidate for electric vehicle (EV) and hybrid electric vehicle (HEV) electric propulsion for its simple and rugged construction, ability of extremely high-speed operation, and insensitivity to high temperatures. However, because SRM construction with doubly salient poles and its non-linear magnetic characteristics, the problems of acoustic noise and torque ripple are more severe than these of other traditional motors. Power electronic technology has made the SRM an attractive choice for many applications. The SRM is a doubly salient, singly excited synchronous motor. The rotor and stator are comprised of stacked iron laminations with copper windings on the stator, as shown in Fig. 5 [4]. The motor is excited with a power electronic inverter that energizes appropriate phases based on shaft position. The excitation of a phase creates a magnetic field that attracts the nearest rotor pole to the excited stator pole in an attempt to minimize the reluctance path through the rotor. The excitation is performed in a sequence that steps the rotor around.

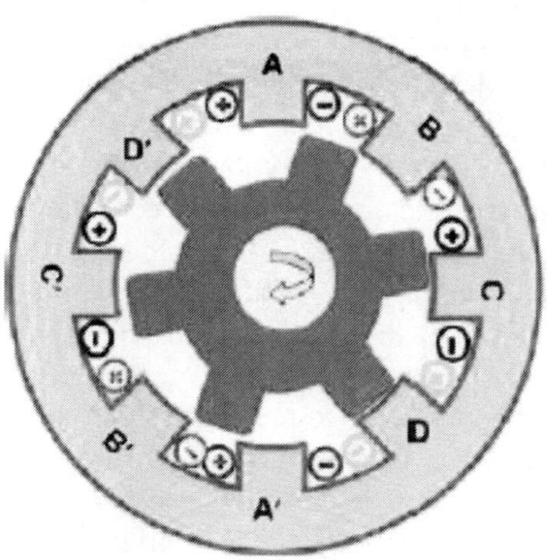

Figure 5. A switched-reluctance motor with 8 stator poles and 6 rotor poles

The SRM is similar in structure to the stepping motor, but it is operated in a manner that allows for smooth rotation. Because there are no permanent magnets or windings on the rotor, all of the torque developed in the SRM is reluctance torque. While the SRM is simple in principle, it is rather difficult to design and develop performance predictions. This is due to the nonlinear magnetic characteristics of the motor under normally saturated operation.

The special design of electric motors used in direct-drive vehicles where the engines are installed in each wheel. This will be discussed more in the mechanical transmission part.

2.3. Supercapacitors Vs. Accumulator Batteries And Fuel Cells

Supercapacitors are relatively new type of capacitors distinguished by phenomenon of electrochemical double-layer, diffusion and large effective area, which leads to extremely large capacitance per unit of geometrical area (in order of multiple times compared to conventional capacitors). They are taking place in the area in-between lead batteries and conventional capacitors. In terms of specific energy (accumulated energy per mass unity or volume) and in terms of specific power (power per mass unity or volume) they take place in the area that covers the order of several magnitudes. Supercapacitors fulfill a very wide area between accumulator batteries and conventional capacitors taking into account specific energy and specific power [1]. Batteries and fuel cells are typical devices of small specific power, while conventional capacitors can have specific power higher than $1MW/dm^3$, but at a very low specific energy. Electrochemical capacitors improve batteries characteristics considering specific power or improve capacitors characteristics considering specific energy in combination with them. In relation to other capacitor types, supercapacitors offer much higher capacitance and specific energies [5-6].

Accumulator batteries and low temperature fuel cells are typical devices with low specific power, where conventional capacitors may have specific power over $1MW/dm^3$, but at very low specific energy. Electrochemical capacitor can improve characteristics of batteries in terms of specific power and improve properties of capacitors in terms of specific energy when they are combined with them [7].

The principal supercapacitor characteristic that makes it suitable for using in energy storage systems (ESS), is the possibility of fast charge and discharge without lost of efficiency, for thousands of cycles. This is because they store electrical energy directly. Supercapacitors can recharge in a very short time having a great facility to supply high and frequent power demand peaks [8].

2.3.1. Supercapacitor Caracterization

Electrochemical investigation methods are widely used for characterization of different kinds of materials, as well as of the processes in systems where the electrochemical reactions take part. There is a series of well known methods, but some new methods from electrotechnical area have been introduced [9-10]. So, first of all it was given an overview of the standard electrochemical methods and parameters, beginning with potential measurement and simple methods such as chronopotentiometry and chronoamperometry, till electrochemical impedance spectroscopy [10]. The last named method is adapted for systems containing large capacitances that became actually with appearance of electrochemical supercapacitors. New methods are Dirac voltage excitation and Dirac current excitation. Measurement system described here allows application of electrochemical methods, as follows: measuring open circuit potential, chronopotentiometry, chronoamperometry, galvanostatic method, potentiostatic method, Dirac voltage excitation, galvanodynamic method, cyclic voltammetry and electrochemical impedance spectroscopy [10-11].

2.3.2. Supercapacitors As A Function Of Increasing Energy Efficiency Of Ev

Most strict requirements are related to supercapacitors applying in electric haulage, i.e. for vehicles of the future. Nowadays, batteries of several hundred farad capacitance are with working voltage of several hundred volts have been produced. Beside great capacitance and relatively high working voltage, these capacitors must have great specific energy and power (because of limited space in vehicle). Considering their specific power, they have great advantage in relation to accumulator batteries, but, on the other side, they are incomparably weaker considering specific energy. Hence, ideal combination is parallel connection of accumulator and condenser batteries. In an established regime (normal drawing) vehicle engine is supplied from accu-battery, and in the case of rapidly speeding, from supercapacitor. Very important is the fact that in the case of abrupt breaking, complete mechanical energy could be taken back to system by converting into electrical energy only in presence of supercapacitor with great specific power [1].

In Figure 6 the scheme of an electrical drive vehicle in which supercapacitor is used for energy storage and so-called regenerative breaking is presented.

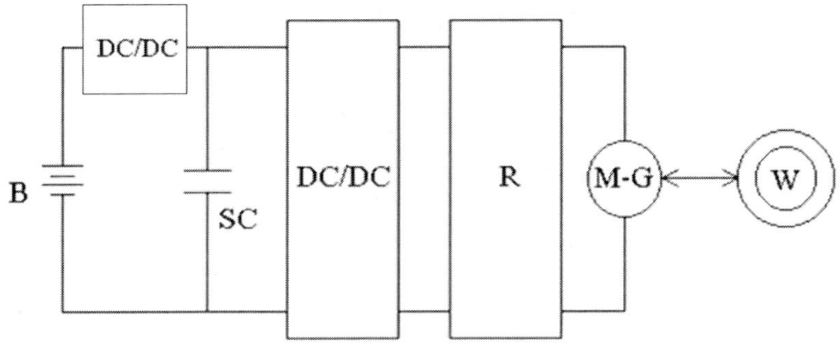

Figure 6. Scheme of electrical drive vehicle with supercapacitor with possibitlity for using breaking energy; B – one-way voltage source, SC – supercapacitor; DC/DC – direct voltage converter; R – regulator; M-G – engine – generator (depending on working regime; W – drive wheels

2.3.3. Supercapacitors In Regulated Electrical Drives

Regulated electrical drives are more than 30% of all electric drives. They are developing quickly and present to constructors stricter and stricter speed regulation (and position) and torque. From energy point of view it is desirable their more participation, since optimal speed setting or required can lead to reduction of used energy [1].

DC source voltage is performed by means of DC-DC converter (chopper). Figure 7 shows principal scheme of such a system.

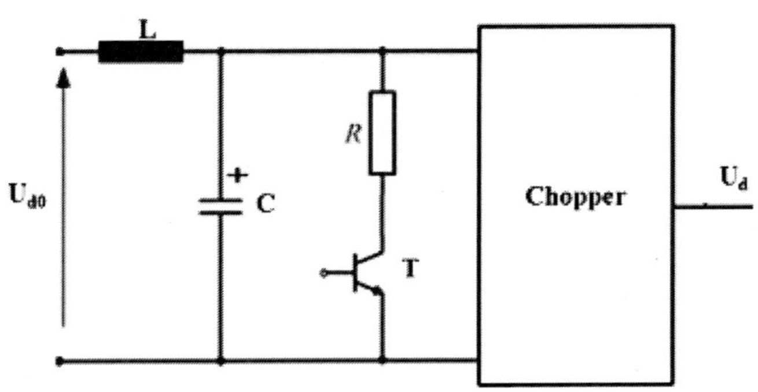

Figure 7. Principal scheme of chopper supply with supercapacitor

To provide breaking, or to dissipate braking energy that cannot be returned to the network through diode rectifier, it is required to have braking device with transistor T and resistor R. Input voltage U_{do} is filtered by simple LC filter and brought to the chopper input that regulates mean value of output voltage U_d.

2.3.4. Accu Batteries

Nowadays, there are a great amount of standard batteries that can be used for EV, however every single type has disadvantages that affect the performance of the vehicle. Therefore, compromises are often made between cost and quality, at the expense of energy efficiency almost all the time. Batteries in combination with supercapacitors are significant improvement and for now this is the system that has the best perspective for future EV.

In the table 1 it is presented the cost per Watt-hour and Specific Energy (Watt-hours per kilogram) for various types of batteries. It is not surprising that the well-known Lead-acid storage batteries head the list. Alkaline cells may be recharged literally dozens of times using the new technology. Recharging alkaline, nickel-cadmium and nickel-metal hydride cells side-by-side in one automatic charger opens up new possibilities for battery selection economy [12].

Table 1. Batteries cost per Watt-hour and Specific Energy

Battery type	Cost, USD/Wh	Specific Energy, Wh/kg
Lead-acid	0.17	41
Alkaline long-life	0.19	110
NiMH	0.99	95
NiCd	1.50	39
Lithium-ion	0.47	128

Costs of lithium-ion batteries are falling rapidly in the race to develop new electric vehicles. The $0.47 price per watt-hour above is for the Nissan Leaf automobile, and they predict a target cost of $0.37 per watt-hour. Tesla

Automobiles uses a smaller battery pack, and they are optimistic about reaching a price of $0.20 per watt-hour in the near future [12].

There is another type of battery that does not appear in the table above, since it is limited in the relative amount of current it can deliver. However, it has even higher energy storage per kilogram, and its temperature range is extreme, from -55 to +150°C. That type is Lithium Thionyl Chloride. It is used in extremely hazardous or critical applications. The specifications for Lithium Thionyl Chloride are $1.16 per watt-hour, 700 Watt-hours per kilogram [12].

Several parameters can be considered for selecting the more adequate battery typology: specific energy, specific power, cost, life, reliability, etc. In addition, it is to be considered that batteries for hybrid electric vehicles require higher powers and lower energies than batteries for pure electric vehicles. Among the previously listed typologies, Lead-acid and Nickel-Cadmium andSodium-Nickel Chloride batteries are normally used on board electric vehicles, because of their low specific powers [13].

2.3.5. Fuel Cells

As far as the fuel cells are concerned, several types are available today, but for vehicle propulsion, Polymer Electrolyte Fuel Cell (PEFC) systems, fed by air and pure hydrogen stored aboard, seem to be highly preferable over other types, mainly because their reduced operating temperature (65-80 degrees depending on the cell design) allow very fast start-up times, and eases the thermal management. A Polymer Electrolyte Fuel Cell is an electrochemical device that converts chemical energy directly into electrical energy, without need of intermediate thermal cycles. It normally consumes H2 and O (typically from Air) as reactants and produce water, electricity and heat. Since cell voltage is so low (less than 1 V), several cells are normally connected in series to form a fuel cell stack with a voltage and power suitable for practical applications.

A fuel cell electric vehicle (FCEV) has higher efficiency and lower emissions compared with the internal combustion engine vehicles. But, the fuel cell has a slow dynamic response. Therefore, a secondary power source is needed during start up and transient conditions. Ultracapacitor can be used as secondary power source. By using ultracapacitor as the secondary power source of the FCEV, the performance and efficiency of the overall system can be improved. In this system, there is a boost converter, which steps up the fuel cell voltage, and a bidirectional DC-DC converter, that couples the ultracapacitor to the DC bus (fig. 8) [13-14].

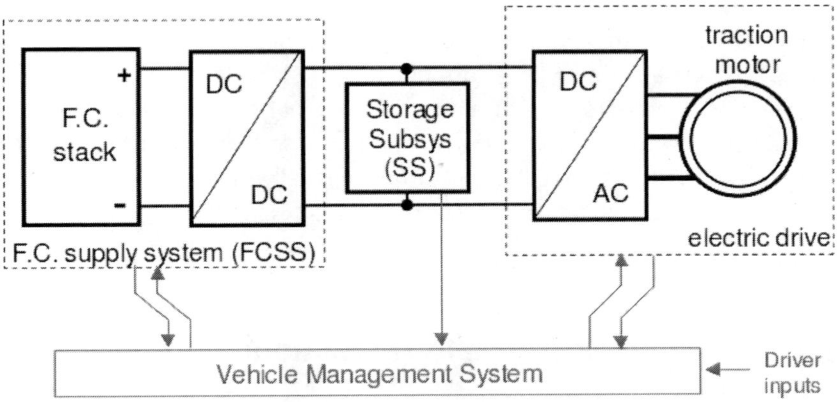

Figure 8. Vehicle with an electrochemical storage system

2.3.6. New Systems

The priority of the EV future development and its commercial success certainly is optimization of the electric power supply. Besides the usual combinations (batteries and supercapacitors, and supercapacitors), researches are going towards new systems that integrate favorable characteristics of the previously used systems.

Typically, standard ultracapacitors can store only about 5% as much energy as lithium-ion batteries. New hybrid system can store about twice as much as standard ultracapacitors, although this is still much less than standard lithium-ion batteries. However, the advantage of ultracapacitors is that they can capture and release energy in seconds, providing a much faster recharge time compared with lithium-ion batteries. In addition, traditional lithium-ion batteries can be recharged only a few hundred times, which is much less than the 20,000 cycles provided by the hybrid system. In other words, the hybrid lithium-ion ultracapacitors have more power than lithium-ion batteries, but less energy storage. In the future, the hybrid lithium-ion ultracapacitor could also be used for regenerative braking in vehicles, especially if it could be scaled up to provide greater energy storage. Since vehicle braking systems need to be recharged hundreds of thousands of times, the hybrid system's cycle life will also need to be improved [15].

Using new processes central to nanotechnology, researchers create millions of identical nanostructures with shapes tailored to transport energy as electrons rapidly to and from very large surface areas where they are stored. Materials behave according to physical laws of nature. The Maryland researchers exploit unusual combinations of these behaviors (called self-assembly, self-limiting reaction, and self-alignment) to construct millions -- and ultimately billions -- of tiny, virtually identical nanostructures to receive, store, and deliver electrical energy [16].

2.4. Reduction Of Losses In The Conductors And Connectors

From the viewpoint of energy efficiency, choice of supply voltage, as well as quality contacts in the connectors and cable section is very important. The designer is limited by other factors such as the security problem (for battery overvoltage), limited space and cost. Therefore, it is necessary to optimize the supply voltage and the conductor section with given constraints. It is similar to the choice of connectors.

Hybrid and electric vehicles have a high voltage battery pack that consists of individual modules and cells organized in series and parallel. A cell is the smallest, packaged form a battery can take and is generally on the order of one to six volts. A module consists of several cells generally connected in either series or parallel. A battery pack is then assembled by connecting modules together, again either in series or parallel [17]. The pack operates at a nominal 375 volts, stores about 56 kilowatt hours (kWh) of electric energy and delivers up to 200 kilowatts of electric power. These power and energy capabilities of the pack make it essential that safety be considered a primary criterion in the pack's design and architecture [18].

Recent battery fires in electric vehicles have prompted automakers to recommend discharging lithium ion batteries following serious crashes. However, completely discharging a vehicle's battery to ensure safety will permanently damage the battery and render it worthless. Self-discharge effects and the parasitic load of battery management system electronics can also irreversibly drain a battery.

Zero-Volt technology relies on manipulating individual electrode potentials within a lithium ion cell to allow deep discharge without inflicting damage to the cell. Quallion has identified three key potentials affecting the Zero-Volt performance of lithium ion batteries. First, the Zero Crossing Potential (ZCP) is the potential of the negative electrode when the battery voltage is zero. Second, the Substrate Dissolution Potential (SDP) is the potential at which the negative electrode substrate begins to corrode. Finally, the Film Dissolution Potential (FDP) is the potential at which the SEI begins to decompose. The crucial design parameter is to configure the negative electrode potential to reach the ZCP before reaching either the SDP or the FDP at the end of discharge. This design prevents damage to the negative electrode which would result in permanent capacity loss. Figure 9 shows a schematic of the voltage profile during deep discharge of Quallion's Zero-Volt cells [18].

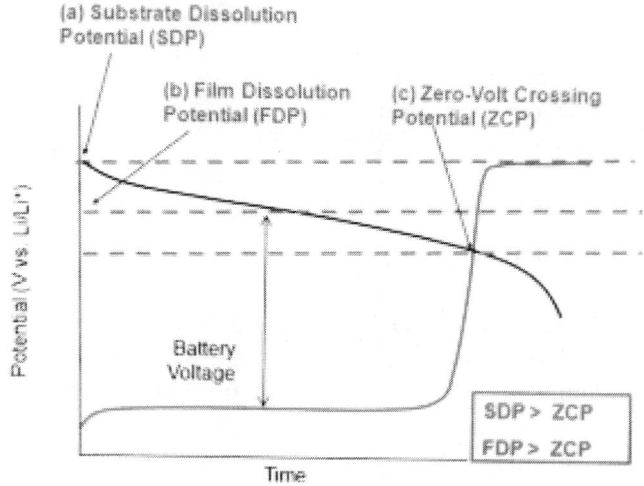

Figure 9. Schematic of key Zero-Volt potentials

Connector contacts are very important, both in terms of energy efficiency (when it comes to high power), and in terms of reliability and security. In recent years, the copper alloy with silver and / or gold is used, but other combinations of metals are to be explored [90.91]. So the compromise between good electrical and mechanical properties, on the one hand, and reasonable prices on the other is required.

Recent literature describes efforts devoted to investigation of copper based alloys in search of improvements in strength and maintenance of strength at high temperatures. The copper-silver alloy is an example of eutectic systems with the eutectic point at 779 °C when the alloy contains 72 % silver and 28 % copper. On both sides of the phase diagram there is a small solubility of the mentioned metals in each other. The maximum solubility of silver in copper is 4.9 at% and the slope of the solvus line indicates the possibility of age-hardening certain alloy compositions. Similar phenomenon - the strengthening of cold worked substitution solid solutions upon annealing up to the re-crystallization temperature is termed anneal hardening. The anneal hardening effect had been observed in Cu-Ag alloys in the annealing temperature range of 140-400 °C, the hardness being increased with the degree of preformation [4,5]. The goal of present work is to investigate corrosion behavior of this alloy obtained by fusion and cast (so called ingot metallurgy - IM) method in different stages of synthesis and thermomechanical treatment. Passivity of copper and its alloys is of interest with respect to basic and applied research due to its wide application in industry. Silver-copper alloys have been investigated elsewhere from the corrosion view point or as an electrode material, but the content of silver in all this alloys overcomes 15 % [19-22].

2.5. Lighting And Heating Of Ev

With the rapid development of high intesive LED technology, it enabled large savings in energy consumption. That fact is crucial for EV. LED and power consumption of exterior vehicle lighting indicated that an all-LED system employing the current generation of LEDs would result in general power savings of about 50% (night time) to about 75% (daytime) over a traditional system. This means that while the long-term fuel cost savings (money) were higher for the gasoline-powered vehicle, long-term distance savings (range) favored the electric vehicle. Now, automotive lighting producer Osram comes to strengthen the idea mentioned above, stating that "micro-hybrids" or mild hybrids, which feature engine stop/start mechanisms to boost the efficiency of conventional vehicles, will benefit greatly from LED lighting by reducing power draw and battery drain, as well as increasing light output during low power mode and startups [23].

Today's roads have very little actual technology incorporated into their design and function. There are many types of technologies which could be incorporated, but we'll begin with what we say is the most important new feature which will soon be applied to actual roads. Since EVs are becoming increasingly popular, while their batteries are still much too weak to assure an anxiety-free drive on the highway, the induction charging (wireless) will begin to be incorporated into one of the lanes, so that these all-electric cars will be able to drive on the highway without using their on-board batteries at all, as they will get their juice straight from underneath the road surface (fig. 10). The idea of inductive charging is simple, and various companies and universities are testing the system now, in view of future mass implementation [23].

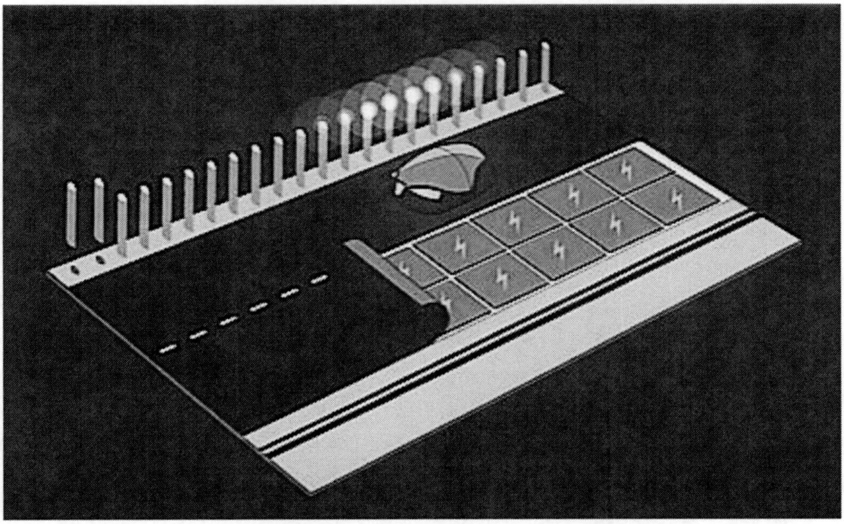

Figure 10. Road surface that charges batteries

Electric vehicles generate very little waste heat and resistance electric heat may have to be used to heat the interior of the vehicle if heat generated from battery charging/discharging can not be used to heat the interior. While heating can be simply provided with an electric resistance heater, higher efficiency and integral cooling can be obtained with a reversible heat pump (this is currently implemented in the hybrid Toyota Prius). Positive

Temperature Coefficient (PTC) junction cooling [24] is also attractive for its simplicity — this kind of system is used for example in the Tesla Roadster.

Some electric cars, for example the Citroën Berlingo Electrique, use an auxiliary heating system (for example gasoline-fueled units manufactured by Webasto or Eberspächer) but sacrifice "green" and "Zero emissions" credentials. Cabin cooling can be augmented with solar power, most simply and effectively by inducting outside air to avoid extreme heat buildup when the vehicle is closed and parked in the sunlight (such cooling mechanisms are available as aftermarket kits for conventional vehicles). Two models of the 2010 Toyota Prius include this feature as an option [25].

3. MECHANICAL LOSSES REDUCTION IN EV

3.1. Tyres Role In Ev

Large impact on the fuel consumption of the cars in general, has tires on its wheels. If the tire optimization is done by the energy efficiency criteria, with acceptable stability, comfort and durability, there is a wide range for development and research. One of the fine examples of the intensive development in this field is racing cars. A modern racing car is a technical masterpiece, but considering the development effort invested in aerodynamics, composite construction and engines it is easy to forget that tyres are still a race car's biggest single performance variable. Average car with good tyres could do well, but it is very known fact that the one with bad tyres, even the very best car did not stand a chance. Despite some genuine technical crossover, race tyres and road tyres are - at best - distant cousins at the moment. An ordinary car tyre is made with heavy steel-belted radial plies and designed for durability - typically a life of 16,000 kilometers or more (10,000 miles). For example, a Formula One tyre is designed to last for, at most, 200 kilometers and it is constructed to be as light and strong as possible. That means an underlying nylon and polyester structure in a complicated weave pattern designed to withstand far larger forces than road car tyres, in [26].

The racing tyre itself is constructed from very soft rubber compounds which offer the best possible grip against the texture of the racetrack, but wear very quickly in the process. All racing tyres work best at relatively high temperatures. For example, the dry 'grooved' tyres used up until very recently were typically designed to function at between 90 degrees Celsius and 110 degrees Celsius [103]. However, electric vehicles can benefit from the years of research and usage of this kind of tyres. The development of the racing tyre came of age with the appearance of 'slick' tyres in the 1960s. Teams and tyre makers realized that, by omitting a tread pattern on dry weather tyres, the surface area of rubber in contact with the road could be maximized. This led to the familiar sight of 'grooved' tyres, the regulations specifying that all tyres had to have four continuous longitudinal grooves at least 2.5 mm deep and spaced 50mm apart. These changes created several new challenges for the tyre manufacturers - most notably ensuring the grooves' integrity, which in turn limited the softness of rubber compounds that could be used, in reference [26].

The 'softness' or 'hardness' of rubber compounds is varied for each road according to the known characteristics of the material that the road was made of. The actual softness of the tyre rubber is varied by changes in the proportions of ingredients added to the rubber, of which the three main ones are carbon, sulfur and oil. Generally speaking, the more oil in a tyre, the softer it will be. Formula One tyres are normally filled with a special, nitrogen-rich air mixture, designed to minimize variations in tyre pressure with temperature. The mixture also retains the pressure longer than normal air would, in [26].

The key characteristics of the new rubber - developed together with the teams in response to the latest aerodynamic regulations - are squarer profiles, increased grip, and softer, more competitive compounds with consistent degradation, optimizing the compounds and profiles to guarantee even better and more stable performance, a longer performance peak, combined with the deliberate degradation that characterized, in [27].

This new measure, which should result in a reduction of aerodynamic down force acting on each tyre, requires a wider and more even contact surface. This objective has been met by having a less rounded shoulder on each tyre and using softer compounds, which produce better grip and more extreme performance, in [27].

Dry weather tyres, known as slicks, are characterized by a tread pattern that is devoid of blocks or channels. Wet weather tyres are characterized by grooves in the tread pattern. The full wet tyres can be easily recognized by the deep grooves in the tread pattern, in reference [28].

At this year's Geneva International Motor Show, one of the tyre manufacturers Goodyear unveils its latest innovation in tyre technology: an extremely low rolling resistance version of its award winning Goodyear EfficientGrip summer tyre with Fuel Saving Technology – specifically developed to fulfill the distinctive requirements of future electric vehicles, in [29]. The look of the tyre inside as well as of the tyre outside is presented in figure 11.

Figure 11. New Goodyear EfficientGrip summer tyre for EV.

The Goodyear EfficientGrip prototype tyre for electric vehicles delivers a range of benefits, including top rated energy efficiency and excellent noise and wet braking performance levels – in combination with Goodyear's latest generation of RunOnFlat Technology for continued mobility after a puncture or complete loss of tire pressure, in [29]. The design of the concept tyre is uniquely suited to complement the performance requirements of electric vehicles. The tyre's narrow shape in combination with its large diameter leads to reduced rolling resistance levels and to a reduced aerodynamic drag and thus reduced energy consumption.

Rolling resistance is mainly caused by the energy loss due to the deformation of the tyre during driving. Less deformation means less energy loss and hence, less rolling resistance. Goodyear engineers used the latest computer simulation technologies to analyze the tyre's potential deformation behavior during driving. The larger rim diameter reduces the overall amount of rubber that is needed, which leads to less rubber deformation during driving. The large tyre diameter requires fewer tire rotations for a certain distance, which in turn results in less heat buildup and tire deformation, which again leads to lower rolling resistance levels and less energy consumption, in [29].

Electric engines often provide a relatively constant torque, even at very low speeds, which increases the acceleration performance of an electric vehicle in comparison to a vehicle with a similar internal combustion engine. This required the development of a modified tread design in combination with a new tread compound to ensure excellent grip especially on dry, and to provide high levels of mileage, in [29].

This EfficientGrip concept tyre showcases our enormous research and development efforts to support the development of electric vehicles with tyres that provide extremely low rolling resistance and noise levels in combination with a very high level of wet performance. Fitted on a standard car this tyre would give 30 percent less rolling resistance which leads to about 6 percent less fuel consumption compared to an average standard summer tyre, in [29].

The effect of tyre pressure on either fuel consumption with regular cars or EV consumption is emphasized. Some researches were done in USA in the last few years. For the control test, the pressure was set at the factory recommended 33 psi in each tire. The subsequent test was done with the pressure set at 45 psi. For each test, the vehicle was driven a total of 550 miles over the course of one week travelling back and forth between the same two cities via the same route. The fuel tank was filled twice per week. Measurement of the quantity of fuel used was taken from the readout on a gas pump at each fill-up. The number of miles travelled was taken from the vehicle's trip odometer, in [190]. Results showed that during the control period, the number of miles travelled per gallon of gasoline consumed was 27. With the tire pressure at 45 psi, the vehicle travelled 30 miles per gallon of gasoline consumed; a difference of 11 percent, in [30].

3.2. Vehicle Body

Automotive design and, specifically, the design of electric and hybrid-electric vehicles, involve a variety of challenges that have to be considered by an appropriate design environment. The convergence of more and more electronics with controls and mechanics makes the design process very complex and involves a variety of technical disciplines. With the complex interactions between the individual system parts, a disconnected consideration of each individual domain is not sufficient anymore. Each individual domain requires specific algorithms and modeling languages to achieve optimal performance for the analysis of that specific domain. A single algorithm usually does not perform for all domains equally; therefore the combination of different algorithms via co-simulation expands the design capabilities of the system considerably (Fig. 12) [31].

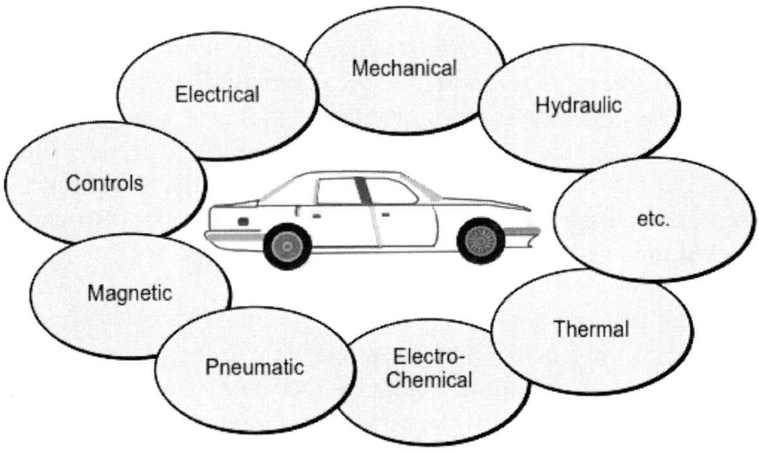

Figure 12. Multi-Domain Design

In recent years simulation programs allow the optimization of vehicle body shapes from the standpoint of energy efficiency. On the other hand, simulations and experiments in the wind tunnel achieve significant energy savings by introducing air turbine, which inevitably airflow into electricity.

3.3. Aerodynamics Of Ev

Moveable aerodynamic components are nothing new, every time you sit on an airliner you see the wing flaps, ailerons moving around, and often as you come into land you can see the array of hydraulics employed to move them. The systems on a Formula 1 racing car work in essentially the same way. Hydraulic tubes, rods and actuators. But whilst on an Airbus A320 or even a modern UAV or fighter jet there is a huge amount of space to work in, on a grand prix car the opposite is true [32]. EV vehicles could benefit a lot from these technologies.

Racing drivers have a new tool at their disposal, called Drag Reduction System (DRS). It is essentially an adjustable rear wing which can be used to facilitate overtaking. The flap is lifted up at the front and pivots about a point at the trailing edge of the wing, so that in the event of a failure, the flap will drop down into the default, high-down force position. Since the timing loops will be sited after corners, drivers will only be able to deploy the active rear wing as a car goes down a particular nominated straight [32].

The materials used in these systems also require great precision. Today in F1 it is mainly titanium tube, though some of what we do involves peek mainly in the fuel system but primarily titanium. Aluminum and stainless steel are also used. Titanium is favored for its inherent lightness and strength, and it means that it is possible to make the cross section of the material so much thinner than if you were using Aluminum. Over the time, the manufacturers have learned to manipulate titanium tubing in ways, especially in small spaces, and the results of that work will be on cars in the future [32].

This week a row has erupted over the design of two teams' diffusers, after the new Williams and Toyota emerged sporting radically different diffuser designs to the other cars launched so far. Williams came up with a 'double decker' diffuser design, while Toyota initially tested an extension to the middle of their diffuser, and then later added a double decker section of their own. Both these designs raised eyebrows up and down the pit lane, as they appear to stretch the wording of the new rules. [33].

As part of the 2009 package of aerodynamic rule changes designed to reduce down force and increase overtaking, the FIA mandated a smaller diffuser in a more rearward position. With the shock of losing 50 per cent of their down force because of these changes, teams have been working hard to get the bodywork shaped to the new rules to regain the lost down force [33].

4. ADDITIONAL ENERGY IN EV

4.1. Solar Cells

Today, world recognizes the synergy between solar panels and electric cars. As the matter a fact there are several car companies that plan to install solar panels in their newer hybrid vehicles. The most important question for most of these manufacturers is: how much extra power will a solar roof panel actually provide? It's very difficult to generate enough power to move a vehicle with energy from the sun's light. So, solar panels at the moment don't have that much of an impact on a hybrid and electric car's efficiency. Solar panels are also made out of silicon, which is too expensive for automakers to use as a viable source [34].

However, there are companies such as Toyota, one of the pioneers in this field, which uses the solar roof panel. Constant technology development will provide better conditions in years that follow for this option. Nowadays, roof panel will power at least part of the hybrid Toyota Prius' air-conditioning unit. Smaller, less power-hungry systems seem to work better with solar power [34].

The most common type of solar panel uses single- or multi-crystalline silicon wafers. Creating the silicon crystal is by far the most energy intensive part of the process, followed by various and sundry manufacturing steps, such as cutting the silicon into wafers, turning the wafers into cells and assembling the cells into modules [35].

The today's electric vehicles consume about 150 watt-hours per kilometer. If the average distance per day is 50 km, then it would be 18,250 kilometers per year. For this calculated consumption, electric vehicle would need to generate 2.75 MWh/year. By this math, monocrystalline solar panels generate about 263 kWh/m^2 per year in the USA. Therefore, about 10.5 square meters of solar panels to completely offset the energy consumed by today's electric vehicles [35]. The only practical place to put panels on the Roadster is the roof (about 1 square meter). Ideally, this would then generate 263 kWh/year. However, the Roadster won't always be in the sun, and it won't be at its ideal angle. A 60% de-rating would be generous to account for shade and suboptimal angles, so the panel would generate about 150 kWh/year – driving the car an additional 3 kilometers per day [35].

However, there is possibility to put solar cells on the other part of the vehicle's surface. The surface from the vehicle's nose, across the hoods, and all the way to the roof can be used for solar cells as presented in figure 13. Also, technology development will without a doubt make progress in increasing solar energy efficiency.

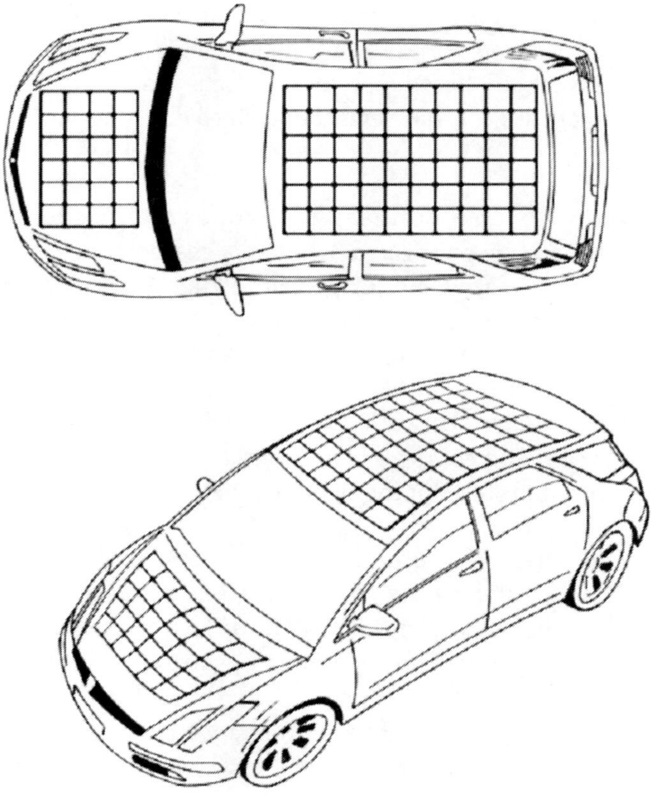

Figure 13. Position of solar cells on the surface of the electric vehicle

4.2. Energy Recovery Systems

4.2.1. Kinetic Energy Recovery Systems

A kinetic energy recovery system (KERS) is an automotive system for recovering a moving vehicle's kinetic energy under braking. The recovered energy is stored in a reservoir (flywheel or a battery or/and supercapacitor)

for later use under acceleration. The device recovers the kinetic energy that is present in the waste heat created by the car's braking process. It stores that energy and converts it into power that can be called upon to boost acceleration, in [36].

The concept of transferring the vehicle's kinetic energy using flywheel energy storage was postulated by physicist Richard Feynman in the 1950s. It is exemplified in complex high end systems such as the Zytek, Flybrid, Torotrak and Xtrac used in F1 and simple, easily manufactured and integrated differential based systems such as the Cambridge Passenger/Commercial Vehicle Kinetic Energy Recovery System (CPC-KERS), in [36].

Xtrac and Flybrid are both licensees of Torotrak's technologies, which employ a small and sophisticated ancillary gearbox incorporating a continously variable transmission (CVT). The CPC-KERS is similar as it also forms part of the driveline assembly. However, the whole mechanism including the flywheel sits entirely in the vehicle's hub (looking like a drum brake). In the CPC-KERS, a differential replaces the CVT and transfers torque between the flywheel, drive wheel and road wheel [36]. KERS Technology is based on a completely new design capable of accumulating power and keeping it in store for the right moment.

KERS Technology works like a turbo charger that provides additional power and acceleration by stiffening the tail of the ski in outturns. The effect: a boost, catapulting the rider into the next turn. Just like when Formula 1 pilots push a button for that extra notch of speed. KERS Technology is an electronic, fully automatic and integrated system. Piezoelectric fibers transform kinetic energy into electrical energy which is stored. Electrical energy is immediately released to areas of the ski, where additional energy is requested. Timing and release are automatically controlled and coordinated. Depending on the flex pattern of different ski models, sensors are programmed beforehand: the more aggressive the ski has to be, the stiffer the tail will become, in reference [37].

The key system features were:

- A flywheel made of steel and carbon fibre that rotated at over 60,000 RPM inside an evacuated chamber
- The flywheel casing featured containment to avoid the escape of any debris in the unlikely event of a flywheel failure

- The flywheel was connected to the transmission of the car on the output side of the gearbox via several fixed ratios, a clutch and the CVT
- 60 kW power transmission in either storage or recovery
- 400 kJ of usable storage (after accounting for internal losses)
- A total system weight of 25 kg
- A total packaging volume of 13 litres, in reference [38]

There are principally two types of system - battery (electrical) and flywheel (mechanical). Electrical systems use a motor-generator incorporated in the car's transmission which converts mechanical energy into electrical energy and vice versa. Once the energy has been harnessed, it is stored in a battery and released when required. Mechanical systems capture braking energy and use it to turn a small flywheel which can spin at up to 80,000 rpm. When extra power is required, the flywheel is connected to the car's rear wheels. In contrast to an electrical KERS, the mechanical energy doesn't change state and is therefore more efficient. There is one other option available - hydraulic KERS, where braking energy is used to accumulate hydraulic pressure which is then sent to the wheels when required, in [36,39].

The first of these systems to be revealed was the Flybrid. This system weighs 24 kg and has an energy capacity of 400 kJ after allowing for internal losses. A maximum power boost of 60 kW (81.6 PS, 80.4 HP) for 6.67 seconds is available. The 240 mm diameter flywheel weighs 5.0 kg and revolves at up to 64,500 rpm. Maximum torque at the flywheel is 18 Nm, and the torque at the gearbox connection is correspondingly higher for the change in speed. The system occupies a volume of 13 liters, in [36].

Nowadays, Formula One has stated that they support responsible solutions to the world's environmental challenges and the FIA allowed the use of 60 kW KERS in the regulations for the 2009 Formula One seasone. Teams began testing systems in 2008: energy can either be stored as mechanical energy (as in a flywheel) or as electrical energy (as in a battery or supercapacitors). As of 2014, in the race cars, the power capacity of the KERS units will increase from 60 kilowatts to 120 kilowatts, in [106].

The aims for introducing KERS technology in the racing world are twofold. Firstly to promote the development of environmentally friendly and road car-relevant technologies in Formula One racing; and secondly to aid overtaking. A chasing driver can use his boost button to help him pass the car in front, while the leading driver can use his boost button to escape. A typical KERS system weighs from 25 to 35 kilograms, in [36,37,39]. For the relevance of the electric vehicles, this energy can be used for supplementing the batteries of electrical engine and thereby adding a few more kilometres to the driving distance at once.

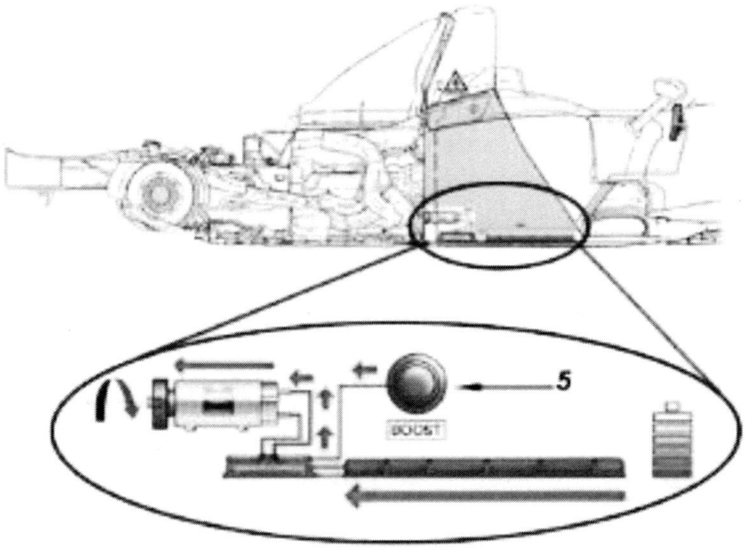

Figure 14. Kinetic Energy Recovery System

Following the current situation, some solutions in KERS packaging has taken a step forwards. Now the energy storage appears to be slightly revised, with the unit inside the gearbox swapped for floor mounted units. The two carbon fiber cases are closed with aluminum tops and are provided with electrical and cooling connections. They sit in the final section of flat floor known as the boat tail, in [40].

Having the units placed on the floor, as opposed to between the gearbox and engine, means they can lower the Centre of Gravity. Also being quite heavy they are placed near the rear axle line to suit the mandatory weight distribution. As mentioned the units are supplied with a common cooling circuit, one pipe routes around the back of the floor to link the devices.

There are also a number of electrical connections for both connecting to the KERS Power Control Unit and for monitoring their status. Quickly detachable connectors are used to allow rapid removal of the floor keeping the units in place, in [40].

The future development appear to have found a new mounting position and format for their KERS energy storage with what appear to be floor mounted super capacitors. Super Capacitors (supercaps) are alternative energy storage to Lithium Ion batteries, using very much the same technology as smaller capacitors used in electronics, in [40].

Typically current F1 cars use dozens of Li-ion cells packed into an array forming a 'battery' pack. This KERS Battery Pack is commonly a single part sat under the fuel tank. Although often used as a single battery, the unit can be broken up into a set of batteries in series. In 2011 Red Bull clearly split this part up into several smaller Battery Packs, there being the two aforementioned units either of the gearbox and another in the gearbox. Although interconnecting these parts with cooling pipes, high current cable and sensor cabling ads some weight, this does provide a nicer packaging solution. It's logical to explain these new floor mounted parts as batteries. However they do not look like the battery packs seen in the gearbox last year, or on other cars. Being on the floor of the car they are subject to even more danger from impacts as well as the heat and vibration that caused issues last year, in [40]. The energy stored in a double-layer capacitor, is used to supply power needed by vehicle electrical systems, in [36].

4.2.2. Waste Heat Energy Recovery

In recent years, there has been active research on exhaust gas waste heat energy recovery for automobiles. Meanwhile, the use of solar energy is also proposed to promote on-board renewable energy and hence to improve their fuel economy. New research in thermoelectric-photovoltaic hybrid energy systems are proposed and implemented for automobiles. The key is to newly develop the power conditioning circuit using maximum power point tracking so that the output power of the proposed hybrid energy system can be maximized. This experimental concept can be easily implemented in electric vehicles [41].

According to the recent studies, General Motors is using shape memory alloys that require as little as a 10°C temperature difference to convert low-grade waste heat into mechanical energy. When a stretched wire made of shape memory alloy is heated, it shrinks back to its pre-stretched length. When the wire cools back down, it becomes more pliable and can revert to its original stretched shape. This expansion and contraction can be used directly as mechanical energy output or used to drive an electric generator. Shape memory alloy heat engines have been around for decades, but the few devices that engineers have built were too complex, required fluid baths, and had insufficient cycle life for practical use. Around 60% of all energy in the U.S. is lost as waste heat; 90% of this waste heat is at temperatures less than 200°C and termed low grade because of the inability of most heat-recovery technologies to operate effectively in this range. The capture of low-grade waste heat, which turns excess thermal energy into useable energy, has the potential to provide consumers with enormous energy savings [42].

For practical use, parts of automotive industry nowadays are working to create a prototype that is practical for commercial applications and capable of operating with either air or fluid based heat sources. GM's shape memory alloy based heat engine is also designed for use in a variety of non-vehicle applications. For example, it can be used to harvest non-vehicle heat sources, such as domestic and industrial waste heat and natural geothermal heat, and in HVAC systems and generators [43].

Thermal Energy Recovery Systems for better fuel efficiency proposes solutions for fuel economy and lower CO2-emissions on combustion engines by making use of their exhaust waste heat. This fuel economy is accessible for engines running on gasoline, diesel, bio fuels, hydrogen or any other type of fuel. This solution proposes high power density for mobile applications and rugged solutions for power generation and marine applications, also being recognized by the motorsport world as an important technology for the future in racing and finally a technology that will contribute to the development of electric vehicle [43].

Plug-in hybrid electric vehicles are already noted for their environmental advantages and fuel savings – but now a new breakthrough technology could mean their fuel economy is boosted by a further seven per cent [44]. Most vehicle waste heat recovery systems that are currently being developed utilize a thermoelectric converter to create electricity, as the name implies, directly from heat. These devices depend on a unique property of certain materials which result in the Seeback effect, discovered in 1821, where the application of heat produces an electric current. The devices have no moving parts. You could think of them as similar to photovoltaic cells, except that they respond to heat rather than light [45].

An effective waste recovery system requires three elements:

1) a thermoelectric material package

2) an electric power management system, which directs the electricity injected into the vehicle's electrical system to the place where it will do the most good at any given time

3) a thermal management system, which is essentially a sophisticated heat exchanger [45]

Some other systems in hybrid electric vehicles reduce fuel consumption by replacing a significant portion of the required electric power normally produced by the alternator with electric power produced from exhaust gas waste heat conversion to electricity in a Thermoelectric Generator Module [46].

4.3. Airflow

It was previously mentioned that vehicle body can be designed to reduce downforce and otherwise adverse airflow. Some of the possibilities are presented here.

During forward motion of an electrically-powered vehicle, air is captured at the front of the vehicle and channeled to one or more turbines. The air from the turbines is discharged at low pressure regions on the sides and/or rear of the vehicle. The motive power of the air rotates the turbines, which are rotatable engaged with a generator to produce electrical energy that is used to recharge batteries that power the vehicle. The generator is rotatable engaged with a flywheel for storing mechanical energy while the vehicle is in forward motion. When the vehicle slows or stops, the flywheel releases its stored energy to the generators, thereby enabling the generator to continue recharging the batteries. The flywheel enables the generators to provide a more stable and continuous current flow for recharging the batteries [47].

It is assumed that the vehicle is moving in a calm and steady wind stream with zero wind velocity. If the vehicle is moving at a constant speed of 15 m/s (54 km/h), then we can think a wind stream with15 m/s is flowing around the vehicle. Normally this wind will cause a drag force which is opposite to the direction of the propulsion of the vehicle. At constant speed (zero acceleration) the energy requirements to move the vehicle forward are –To overcome the frictional force (rolling resistance of road) and to overcome wind resistance [48]. At this Condition, if the air stream flowing around the vehicle (which was not interacting with the vehicle previously) is allowed to enter inside and let it flow down to the rear side; then it may be possible to use these air streams to generate power. The vehicle has already interacted with this wind and it d eflects the stream of wind at the two sides of it by stagnation at the front.

This is the energy that had been lost from the vehicle to overcome the aerodynamic resistant. Now if these stream generated by the interaction of the wind and vehicle is captured within the vehicle in such a way that it would not impose an additional drag at the direction of propulsion of the vehicle, some of the energy can be recovered and fed back to the battery by means of conventional energy conversion processes. Placing a wind turbine can serve the purpose. At the same time it will help to increase the pressure at the back side (according to Bernoulli's equation pressure will be increased if velocity is decreased and velocity will be reduced at the back side of the turbine after energy extraction) which will reduce the drag force that existed before with the conventional design of the vehicle. So, vortex shedding will be reduced at the rear side. For this it is necessary to modify the design of a vehicle which gives provision of air flow through the vehicle. On the other hand positioning of the turbines will also be important because they must be placed in such a way that they do not impose or create any additional drag on the vehicle. Symmetrical positioning of the turbine can do t he trick as the thrust acting on the turbines will cancel each other (Fig. 15) [49].

4.4. Hybrid Electric Vehicle

Generally, hybrid vehicles could be described as vehicles using combination of technologies for energy production and storage. Two types of the vehicles are in consideration – so called parallel and linear hybrids. Parallel type possesses mechanical connection between power generator and drive wheels, while in linear one such connection does not exist. Serial hybrids have significant advantages in relation to parallel ones because of their mechanical simplicity, design flexibility and possibility for simple incorporation of new technologies [1].

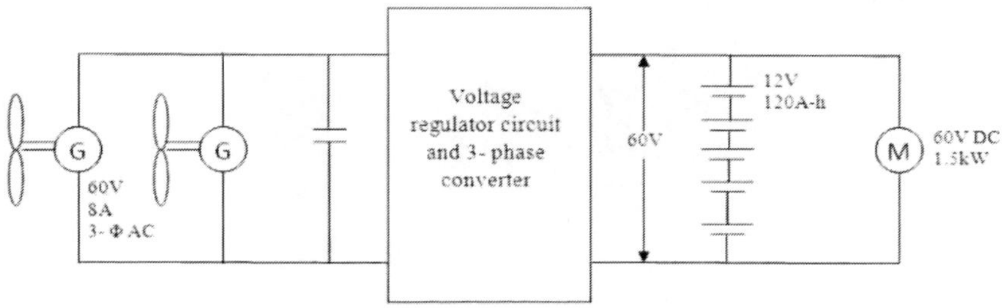

Figure 15. Charging and control circuit of the battery

Hybrid electric vehicles (HEVs) combine the internal combustion engine of a conventional vehicle with the high-voltage battery and electric motor of an electric vehicle. As a result, HEVs can achieve twice the fuel economy of conventional vehicles (Fig. 1). In combination, these attributes offer consumers the extended range and rapid refueling they expect from a conventional vehicle, as well as much of the energy and environmental benefits of an electric vehicle. HEVs are inherently flexible, so they can be used in a wide range of applications — from personal transportation to commercial hauling. Hybrid electric vehicles have several advantages over conventional vehicles:

- Greater operating efficiency because HEVs use regenerative braking, which helps to minimize energy loss and recover the energy used to slow down or stop a vehicle;

- Lighter engines because HEV engines can be sized to accommodate average load, not peak load, which reduces the engine's weight;

- Greater fuel efficiency because hybrids consume significantly less fuel than vehicles powered by gasoline alone;

- Cleaner operation because HEVs can run on alternative fuels (which have lower emissions), thereby decreasing our dependency on fossil fuels (which helps ensure our national security); and

- Lighter vehicle weight overall because special lightweight materials are used in their manufacture.

Hybrid electric vehicles are becoming cost-competitive with similar conventional vehicles, and most of the cost premium can be offset by overall fuel savings and tax incentives. Some states even offer incentives to consumers buying HEVs [50].

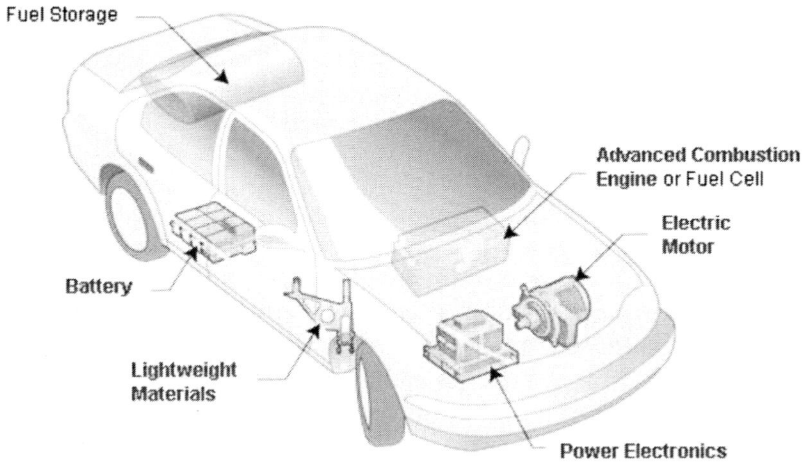

Figure 16. HEV

4.5. Today's High-Speed Ev

Nowadays, the most powerful high-performance electric vehicle has four electric motors producing a total output of 552 kW and a maximum torque of 1000 Nm. As a result, the gullwing model has become the world's fastest electrically-powered series production vehicle accelerates from zero to 100 km/h in 3.9 seconds [51].

Enormous thrust comes courtesy of four synchronous electric motors providing a combined maximum output of 552 kW and maximum torque of 1000 Nm. The very special gullwing model accelerates from zero to 100 km/h in 3.9 seconds, and can reach a top speed of 250 km/h (electronically limited). The agile response to accelerator pedal input and the linear power output provide pure excitement: unlike with a combustion engine, the build-up of torque is instantaneous with electric motors – maximum torque is effectively available from a standstill. The spontaneous build-up of torque and the forceful power delivery without any interruption of tractive power are combined with completely vibration-free engine running characteristics [51].

The four compact permanent-magnet synchronous electric motors, each weighing 45 kg, achieve a maximum individual speed of 13,000 rpm and in each case drive the 4 wheels selectively via a axially-arranged transmission design. This enables the unique distribution of torque to individual wheels, which would normally only be possible with wheel hub motors which have the disadvantage of generating considerable unsprung masses [51].

Battery efficiency, performance and weight are by far the most important factors in electric vehicles. The high-voltage battery in the current high-performance electric vehicles boasts an energy content of 60 kWh, an electric load potential of 600 kW and weighs 548 kg – all of which are absolute best values in the automotive sector. The liquid-cooled lithium-ion high-voltage battery features a modular design and a maximum voltage of 400 V. Advanced technology and know-how from the world of Formula 1 have been called on during both the development and production stages [51].

The high-voltage battery consists of 12 modules each comprising 72 lithium-ion cells. This optimized arrangement of a total of 864 cells has benefits not only in terms of best use of the installation space, but also in terms of performance. One technical feature is the intelligent parallel circuit of the individual battery modules – this helps to maximize the safety, reliability and service life of the battery. As in Formula 1, the battery is charged by means of targeted recuperation during deceleration whilst the car is being driven [51].

A high-performance electronic control system converts the direct current from the high-voltage battery into the three-phase alternating current which is required for the synchronous motors and regulates the energy flow for all operating conditions. Two low-temperature cooling circuits ensure that the four electric motors and the power electronics are maintained at an even operating temperature. A separate low-temperature circuit is responsible for cooling the high-voltage lithium-ion battery. In low external temperatures, the battery is quickly brought up to optimum operating temperature with the aid of an electric heating element. In extremely high external temperatures, the cooling circuit for the battery can be additionally boosted with the aid of the air conditioning. This also helps to preserve the overall service life of the battery system [51].

Ideally the EV is charged with the aid of wall box. As it could be installed in a home garage, this technology provides a 22 kW quick-charge function, which is the same as the charging performance available at a public charging station. A high-voltage power cable is used to connect the vehicle to the wall box, and enables charging to take place in around three hours. Charging takes around 20 hours without the wall box [51].

To ensure maximum safety, the SLS AMG Coupé Electric Drive, one of the most advanced high-performance EV today, makes use of an eight-stage safety design. This comprises the following features:

- All high-voltage cables are color-coded in orange to prevent confusion
- Comprehensive contact protection for the entire high-voltage system
- The lithium-ion battery is liquid-cooled and accommodated in high-strength aluminium housing within the carbon-fibre zero-intrusion cell
- Conductive separation of the high-voltage and low-voltage networks within the vehicle and integration of an interlock switch
- Active and passive discharging of the high-voltage system when the ignition is switched to "off"
- In the event of an accident, the high-voltage system is switched off within fractions of a second
- Continuous monitoring of the high-voltage system for short circuits with potential compensation and insulation monitors
- Redundant monitoring function for the all-wheel drive system with torque control for individual wheels, via several control units using a variety of software

By using this design, EV manufacturers ensures maximum safety during production of the vehicle and also during maintenance and repair work [51].

The intelligent and permanent all-wheel drive concept, with four motors for four wheels guarantees driving dynamics at the highest level, while at the same time providing the best possible active safety. Optimum traction of the four driven wheels is therefore ensured, whatever the weather conditions. According to the developers, the term "Torque Dynamics" refers to individual control of the electric motors, something which enables completely new levels of freedom to be achieved. The AMG Torque Dynamics feature is permanently active and allows for selective distribution of forces for each individual wheel. The intelligent distribution of drive torque greatly benefits driving dynamics, handling, driving safety and ride comfort. Each individual wheel can be both electrically driven and electrically braked, depending on the driving conditions, thus helping to:

- optimize the vehicle's cornering properties
- reduce the tendency to over steer/under steer
- increase the yaw damping of the basic vehicle
- reduce the steering effort and steering angle required
- increase traction

AMG Torque Dynamics system enables optimum use of the adhesion potential between the tires and the road surface in all driving conditions. The technology allows maximum levels of freedom and as such optimum use of the critical limits of the vehicle's driving dynamics [51].

The trailblazing body shell structure of the SLS AMG Coupé Electric Drive is part of the ambitious "AMG Lightweight Performance" design strategy. The battery is located within a carbon-fiber monocoque which forms an integral part of the gullwing model and acts as its "spine". The monocoque housing is firmly bolted and bonded to the aluminum space frame body. The fiber composite materials have their roots in the world of Formula 1, among other areas. The advantages of CFRP (carbon-fiber reinforced plastic) were exploited by the Mercedes-AMG engineers in the design of the monocoque. These include their high strength, which makes it possible to create extremely rigid structures in terms of torsion and bending, excellent crash performance and low weight. Carbon-fiber components are up to 50 percent lighter than comparable steel ones, yet retain the same level of stability. Compared with aluminum, the weight saving is still around 30 percent, while the material is considerably thinner. The weight advantages achieved through the carbon-fiber battery monocoque are reflected in the agility of the electric vehicle and, in conjunction with the wheel-selective four-wheel drive system, ensure true driving enjoyment. The carbon-fiber battery monocoque is, in addition, conceived as a "zero intrusion cell" in order to meet the very highest expectations in terms of crash safety. It protects the battery modules inside the vehicle from deformation or damage in the event of a crash [51].

The basis for CFRP construction is provided by fine carbon fibers, ten times thinner than a human hair. A length of this innovative fiber reaching from here to the moon would weigh a mere 25 grams. Between 1000 and 24,000 of these fibers are used to form individual strands [51].

The purely electric drive system was factored into the equation as early as the concept phase when the super sports car was being developed. It is ideally packaged for the integration of the high-performance, zero-emission technology: by way of example, the four electric motors and the two transmissions can be positioned as close to the four wheels as possible and very low down in the vehicle. The same applies to the modular high-voltage battery. Advantages of this solution include the vehicle's low center of gravity and balanced weight distribution – ideal conditions for optimum handling, which the electrically-powered gullwing model shares with its petrol-driven sister model. Another distinguishing feature is the speed-sensitive power steering with rack-and-pinion steering gear: the power assistance is implemented electro hydraulically rather than just hydraulically [51].

The high-performance ceramic composite brakes are used in the latest electrical vehicles, which boast direct brake response, a precise actuation point and outstanding fade resistance, even in extreme operating conditions. The over-sized discs – measuring 402 x 39 mm at the front and 360 x 32 mm at the rear – are made of carbon fiber-strengthened ceramic, feature an integral design all round and are connected to an aluminum bowl in a radially floating arrangement. The ceramic brake discs are 40 percent lighter in weight than the conventional, grey cast iron brake discs. The reduction in unsprung masses not only improves handling dynamics and agility, but also rides comfort and tire grip. The lower rotating masses at the front axle also ensure a more direct steering response – which is particularly noticeable when taking motorway bends at high speed [51].

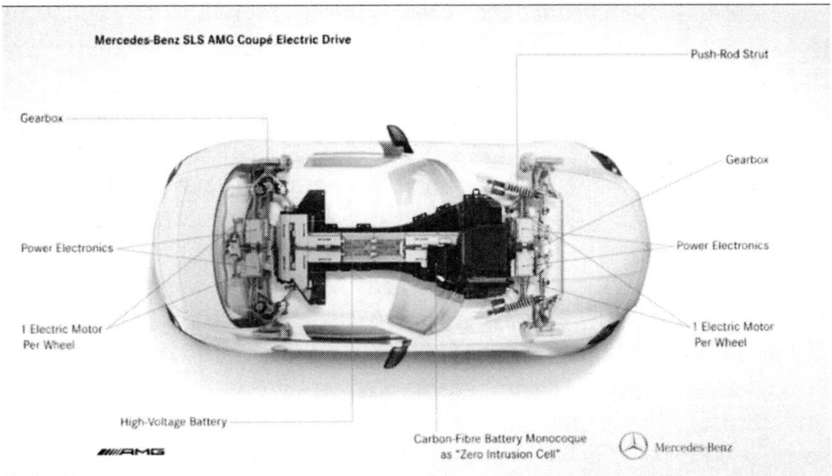

Figure 17. Today's high-speed EV system

5. DRIVING OPTIMIZATION

5.1. Comfort, Information And Safety

Minimizing electricity consumption is often in conflict with comfort and even security of vehicles and people. That's why new technologies are being used to increase safety and comfort, and still energy consumption to be on a low level. Some of the current opportunities and trends are presented in here.

5.1.1. Computer Control

Nowadays, computers are indispensable part of every vehicle. It monitors and controls virtually all vehicle functions, but also processed and displayed a lot of additional information, which significantly contributes to the comfort and safety. In EV that trend is particularly used. The vehicle is equipped with sensors that provide input data and further processed in a computer. The obtained results act on actuators, or the situation is shown on the display and the decision is left to the man [52-53].

Sensors are elements that receive and convert non-electrical signals into electrical. Temperature shift (translation, rotation, stretching), pressure, brightness, electromagnetic radiation, magnetic fields can be detected and can be converted. The temperature is the most usually measured as the non-electrical input, therefore many types of sensors are developed over the years. There are NTC (Negative Temperature Coefficient) and PTC (Positive Temperature Coefficient) resistors and thermocouples [52-54].

In modern vehicle, for the measurement of ambient temperature, cabin and equipment itself, semiconductor sensors are used. They are the product of modern technology of silicon (Si) integrated circuits, therefore also called Si sensors. Silicon sensors consist of integrated circuits using temperature-active properties of semiconductor compounds. All sensors can be with current or voltage output. In both cases, the output signal is proportional to the absolute temperature. The amplitude of the output signal is relatively high and linear, and the interpretation of the signals can be done without any difficulties. Si sensors temperature range usually is from -50 ° C to +150 ° C. The stability and accuracy of these sensors is good enough to allow readings with ± 0.1 ° C resolution. Thermal imager is used for more complex state visual monitoring used the [55].

For the measurement of other important physical quantities (pressure, force, position, displacement and level), sensors that respond to physical movement and / or movement are used. The most commonly used types are semiconductors and resistant strain gauges, linear voltage displacement transducers (LVDT), resistive potentiometers and capacitive sensors. Although each of these sensors is based on different principles, the output signals of all the sensors are voltage, current and impedance. These signals are directly or indirectly analog voltage expressed, so all the techniques described for the measurement are related to these transducers. Sensors that require external excitation reduce the accuracy of the measurement. Higher excitation levels provide higher levels of the output. However, the higher excitation increases internal power dissipation and measurement error, even with mechanical transducers. Each transducer has its own optimal level of excitation [54].

Flow and velocity quantities are measured using resistive, piezoelectric, thermal, and other transducers. As mentioned earlier, all methods ultimately provide as output an analog voltage, current, or impedance. Types of transducers, such as rotary encoders, turbine, magnetic and optical sensors, have digital or pulse outputs. Speed or number of events can be determined by using digital counters and frequency meter [54,56,57].

Two-way communication between humans and computers is done through the touch screen display. Touch screen allows user to interact with a computer through touching the mark and the image on the screen. It is a visual electronic device that can sense touch and determine its location on the surface. The touch itself means contact between human fingers and the screen. The touch screen can also register contact other passive objects, such as special pens, styluses (used for greater precision and less contaminating the screen). Ability to register touch on the touch screen display depends on the implemented touch technology: ones can register just one touch and its position at a given time (single touch), others are capable of registering two or more simultaneous touch and their position on the screen (multi-touch).

Touch screen displays eliminate constraints on a number of discrete keys that are present in conventional membrane keypad. With a touch screen, combined with digital high-resolution display and integrated software [58-59], now there are virtually millions of switching options available for the user.

In commercial terms, touch screen displays, as devices with touch technology, make computer technology easy to use and accessible to all and also significantly to reduce time and cost of training of its use. They also provide much faster access to information as touch technology simplifies and speeds up the search process, which is crucial to driving. As an assembly that is mounted in front of a video display, touch screen display has an independent XY coordinate system that is calibrated according to the matrix display. To determine the location of the touch in the simplest implementation it requires two measurements, one to determine the coordinates of the X-axis and one to determine the coordinates of the Y axis. These measurements are then converted to the coordinates of the point of contact, which is then sent to the host (PC or microcontroller) via serial communication port [60]. A typical example of the application of TS and microprocessor technology is a GPS navigation system (Fig. 18) [61].

Figure 18. GPS navigation system

Figure 19. Complete computer in a hermetically closed housing

An example of a complete computer in a hermetically closed housing is shown in Fig. 19. Nexcom Company has released transport intended fanless computer - VTC-3300, for vehicles and fleet management [62].

5.1.2. Fire Protection In Ev

EV and HEV in particular have a lot of critical areas where it can get to the inception of fire. This requires a vehicle equipped elements of fire protection.

Central unit for fire detection and fire alarm, or as it is often called central unit for fire detection and its task is to power supply detectors and detection lines with stable and regulated supply voltage, which should be available in all expected operational situations, able to take a normal signal status, alarm status, interception line signal or removing the detector signal, short circuit signal, to signalize received state at the central unit and to forward signal to the sound and light devices and to ensure that the executive functions of the system that are required. Alarm indication at the control unit can turn on the respective light emitting diodes, or additional information through the display, but also by activating an internal audible alarm, buzzer or horn. Today, the central fire protection unit connects to the computer, or it is incorporated as software in the computer system.

In addition to the central unit detection system must include detectors, alarms and detection and alarm lines, also the connections to the device that activates the sound and / or light alarms and executive functions. Network that connects the detection system elements is performed mainly by cables and its careful design and selection are essential to the quality, safety and value rationality of the system.

While in conventional systems alarm identification is with group of detectors, central unit and the person that receives information about the group (zone) that alarm is on, however with addressable system each detector gets its code (address) that identifies and tells to the central unit and to the present stuff its state. So the group identification alarm systems, the central unit receive information from a group of detectors (zones) of the alarm or some other event. Event means any change of state of the zone, such as an alarm, a signal failure, signal extraction detectors, fine lines and so on. Whereas in these systems, after the alarm of any zone a person in charge comes to the site, review of the protected object and determines the place where the alarm originated, the addressable system has been known for the receipt of alarms and place of origin, the detector that is in alarm state and the place where it is placed [54].

5.1.3. Ir Termography

IR thermography is a highly sophisticated measuring technique whose beginnings of extensive using coincide with the beginning of the third millennium. The reasons for this can be found in the fact that the thermography camera as a device that provides a visualization of thermal radiation would have to be consisting of many of the latest developments in science so that they become commercially available and easy to use. Very high breakthroughs in the field of sensors, thin films, optoelectronics, microelectronics and microcomputers are integrated and incorporated into these modern devices adapting them to the requirements of users in almost all areas of human activity.

The word thermography (literally, it would mean see the heat) explains the essence of this concept. Specifically, the point is that the appropriate devices (cameras) translate waves from the infrared region into a selected color of the visible part of the electromagnetic spectrum making them visible to the human eye. Different temperatures at the same time correspond to different colors and shades of colors and it is possible even to choose the color palette in which we want to show the resulting temperature map of the object [55, 63,64]. In modern vehicles is incorporated one or more thermography imagers and monitors the state of driver, equipment, or danger on the road in case of low visibility.

Camera which recorded persons in certain position is fixed, for example in front of mirror. Functionality of system is observed and recorded images were compared with literature data. Especially, it was taken care of record conditions: day-time record, time of taking drugs or active substance (coffee, alcohol, tea), room temperature, personal conditions such as emotions, satiety, hunger and physical activity. It was found that in normal conditions temperature in ocular region of healthy person does not exceed 36.3°C. In case of fever it is significantly higher. Thermograms of healthy person before and after vigorous physical activities show also the temperature changes in means of increase (fig. 20)[65].

The fact that thermography can detect very small differences in temperature gives the ability to detect the presence of persons (fig. 21), or animals (fig. 22) at night or in conditions of dense fog. Thermal detectors can function in the complete absence of any light. This makes them the perfect tool for observation in absolute darkness. Potential danger on the road in such conditions can be detected at a distance of 400 m for some systems, up to several kilometers, depending on the equipment and requirements.

Figure 20. Thermogram of driver

Figure 21. Thermogram of persons on the road

Figure 22. Thermogram of animal on the road

5.2. Route

Route optimization (RO) is an important feature of the Electric Vehicles which is responsible for finding optimized paths between any source and destination nodes in the road network. Recent researches perform the RO for EV using the Multi Constrained Optimal Path (MCOP) problem. The proposed MCOP problem aims to minimize the length of the path and meets constraints on total travelling time, total time delay due to signals, total recharging time, and total recharging cost. The proposed algorithms need to have innovative methods for finding the velocity of the particles and updating their positions with accurate database of the requested roads [66-67].

6. CONCLUSION

Electric drive vehicles are one of the most advanced vehicles at the moment taking into account contamination of environment. Lately there is an increased interest in the world for hybrid vehicles that have smaller fuel consumption and substantially less contamination emission footprint. Hybrid vehicles in most general terms can be described as vehicles comprising combination of energy producing and storing.

In this chapter, possibilities of energy savings in EV and HEV, energy generating in the vehicle itself and measures to improve comfort and safety are presented.

Therefore they must be combined with super capacitors. Beside the development of standard technologies, development of power supply is crucial for EV. Accumulator batteries and fuel cells still have not reached the level to obscure enough for autonomy and meet the dynamic characteristics of vehicles. Super capacitors are only available technology today that can provide high power and great cycle numbers at acceptable price. Super capacitors have other properties that makes them interesting in hybrid vehicles, and it's ability of complete regeneration of energy of braking (so called regenerative braking), which increases energy efficiency, no special maintenance needed, great utilization of electric energy, small toxicity and easy storage after use.

ACKNOWLEDGEMENTS

This work was financially supported by the Ministry of Science and Technological Development Republic of Serbia (Projects No. 172060 and TR 32043).

REFERENCES

1. Zoran Stević, Mirjana Rajčić-Vujasinović, Supercapacitors as a Power Source in Electrical Vehicles, Book title: Electric Vehicles – The Benefits and Barriers / Book 1, Edited by: Seref Soylu, Intech, Rijeka (2011)

2. Miroslav Bjekić, Zoran Stević, Alenka Milovanović, Sanja Antić, Regulacija elektromotornih pogona, Tehnički fakultet, Čačak (2010)

3. Dragan Milivojević, Zoran Stević and Mirjana Rajčić-Vujasinović, Hardware and Software of a Bipolar Current Source Controlled by PC, Sensors 8 (2008) 1977- 1983

4. Omar Ellabban, Review of Torque Control Strategies for Switched Reluctance Motor in Automotive Applications, *ieeeexplore.net/search/*

5. Z. Stević, M. Rajčić-Vujasinović, Chalcocite as a potential material for supercapacitors, Journal of Power Sources 160 (2006) 1511-1517

6. Zoran Stević, M. Rajčić-Vujasinović, S. Bugarinović and A.Dekanski, Construction and Characterisation of Double Layer Capacitors, Acta Physica Polonica A, Vol. 117 (2010)1, 228-233

7. Stević, Z. Supercapacitors based on copper sulfides, Ph.D. Thesis, University of Belgrade, 2001.

8. Arbizzani, C.; Mastragostino, M. & Soavi, F. (2001). New trends in electrochemical supercapacitors. *Journal of Power Sources,* Vol.100, No1-2, (November 2001), pp. 164-170.

9. M. Rajcic-Vujasinovic, Z.Stankovic, Z. Stevic, The consideration of the electrical circuit analogous to the copper or coppersulfide/electrolyte interfaces based on the time transient analysis, Russian Journal of Electrochemistry 35,3 (1999) 347- 354

10. Zoran Stević, Mirjana Rajčić Vujasinović, Aleksandar Dekanski, Estimation of Parameters Obtained by Electrochemical Impedance Spectroscopy on Systems Containing High Capacities, Sensors 9 (2009) 7365-7373

11. Zoran Stević, Zoran Andjelković, Dejan Antić, A New PC and LabVIEW Package Based System for Electrochemical Investigations, Sensors 8 (2008) 1819-1831

12. http://www.allaboutbatteries.com/Battery-Energy.html

13. Carmine Miulli, Testing Modeling and Simulation of Electrochemical Systems, Doctoral Thesis, University of Pisa

14. Samosir, A.S.; Yatim, A.;, "Dynamic evolution control of bidirectional DC-DC converter for interfacing ultracapacitor energy storage to Fuel Cell Electric Vehicle system," *Power Engineering Conference, 2008. AUPEC '08. Australasian Universities,* vol., no., pp.1-6, 14-17 Dec. 2008

15. http://phys.org/

16. www.nanocenter.umd.edu

17. MIT University, USA http://mit.edu/ evt/summary_battery_ specifications.pdf

18. http://www.batterypoweronline.com

19. Grekulović Vesna J., Rajčić-Vujasinović Mirjana M., Stević Zoran M., Electrochemical behavior of Ag-Cu alloy in alkaline media, Chemical Industry, 64 (2) (2010) 105-110

20. Mirjana Rajčić-Vujasinović, Svetlana Nestorović, Vesna Grekulović, Ivana Marković, Zoran Stević, Electrochemical behavior of cast CuAg4at% alloy, Corrosion, 66(2010)10, 105004-1-5

21. Mirjana Rajčić-Vujasinović, Svetlana Nestorović, Vesna Grekulović, Ivana Marković, Zoran Stević, Electrochemical behaviour of sintered CuAg4at.% alloy, Metallurgical and Materials Transactions B, 41B(2010)5, 955-961

22. Vesna Grekulović, Mirjana Rajčić-Vujasinović, Batrić Pešić, Zoran Stević, Influence of BTA on Electrochemical Behavior of AgCu50 Alloy, Int. J. Electrochem. Sci., 7 (2012) 5231 - 5245.

23. http://www.autoevolution.com

24. US Patent, US 5889260, Golan, Gad & Yuly Galperin, "Electrical PTC heating device"

25. "2010 Options and Packages". *Toyota Prius*. Toyota

26. Formula One official site, Report http:// www.formula1.com /inside_f1/understanding_the_sport/5283.html

27. Formula One official site, Report http:// www.formula1.com/news /headlines/2012/1/12953.html

28. Pirelli official site, Report, http://www.pirelli.com/tyre/ww/en/f1/tyre-range.html

29. Goodyear official site, Report http://www.goodyear.eu/uk_en/news/ 80191-goodyear_tire_for_electric_vehicles_delivers

30. http://thesmartdrive.com/2009/07/the-effect-of-tire-pressure-on-fuel-economy/

31. Dr. Uwe Knorr, Dr. Ralf Juchem, A complete co-simulation-based design environment for electric and hybrid-electric vehicles, fuel-cell systems, and drive trains, Ansoft Corporation, Pittsburgh, PA

32. http://www.racecar-engineering.com/articles/f1/drs-the-drag-reduction-system/

33. http://www.autosport.com/news/report.php/id/73083

34. http://auto.howstuffworks.com/fuel-efficiency/hybrid-technology/hybrid-cars-utilize-solar-power1.htm

35. http://www.teslamotors.com/blog/electric-cars-and-photovoltaic-solar-cells

36. http://en.wikipedia.org/wiki/Kinetic_energy_recovery_system

37. http://www.head.com/ski/technologies/skis/?region=eu&id=330

38. http://www.flybridsystems.com/F1System.html

39. http://www.formula1.com/inside_f1/understanding_the_sport/8763.html

40. http://scarbsf1.wordpress.com/2012/04/17/red-bull-kers-floor-mounted-super-capacitors/

41. Xiaodong Zhang; Chau, K.T.; Chan, C.C.; Gao, S.;, "An automotive thermoelectric-photovoltaic hybrid energy system," Vehicle Power and Propulsion Conference (VPPC), 2010 IEEE, vol., no., pp.1-5, 1-3 Sept. 2010

42. http://www.heat2power.net

43. http://www.f1technical.net/forum/viewtopic.php?f=4&t=5751

44. http://www.thegreencarwebsite.co.uk/blog/index.php/2012/10/12/huge-breakthrough-for-plug-in-hybrid-electric-vehicles/

45. http://www.triplepundit.com/2010/07/vecarious-will-turn-you-car%E2%80%99s-waste-heat-into-energy/

46. LaGrandeur, J.; Crane, D.; Hung, S.; Mazar, B.; Eder, A.;, "Automotive Waste Heat Conversion to Electric Power using Skutterudite, TAGS, PbTe and BiTe," *Thermoelectrics, 2006. ICT '06. 25th International Conference on*, vol., no., pp.343-348, 6-10 Aug. 2006 US Patent 5,680,032

47. S.M. Ferdous, Walid Bin Khaled, Benozir Ahmed, Sayedus Salehin, Enaiyat Ghani Ovy, Electric Vehicle with Charging Facility in Motion using Wind Energy, World Renewable Energy Congress 2011, Sweden S.M. Ferdous, Walid Bin Khaled, Benozir Ahmed, Sayedus Salehin, Enaiyat Ghani Ovy, World Renewable Energy Congress 2011-Sweden

48. http://www1.eere.energy.gov/vehiclesandfuels

49. http://www.supercars.net/cars/5835.html Story by Daimler AG

50. I. Radovanovic, I. Popovic, *"Implementation of Smart Transducer Correction Functions"*, YUINFO2012 Conference, Feb 2012

51. N. Bezanic, I. Popovic, I. Radovanovic, *"Implementation of Service Oriented Architecture in Smart Transducers Network"*, YUINFO2012 Conference, Feb 2012

52. Zoran Stević, Optoelektronics, University of Belgrade, Technical faculty in Bor (2005)

53. Zoran Stević, Mirjana Rajčić-Vujasinović, Dejan Antić, Thermovision applications, University of Belgrade, Technical faculty in Bor (2008)

54. I. Radovanovic, N. Rajovic, V. Rajovic, J. Jovicic, *"Signal Acquisition And Processing in the Magnetic Defectoscopy of Steel Wire Ropes"*, TELFOR Conference, Nov 2011

55. I. Radovanovic, N. Rajovic, V. Rajovic, J. Jovicic, *"Signal Acquisition And Processing in the Magnetic Defectoscopy of Steel Wire Ropes"*, TELFOR Journal, Vol. 4, No. 2, 2012

56. Dj. Klisic, M. Zlatanovic,I. Radovanovic, *"Application database of wind measuring stations for testing CFD model of software development tool WindSim"*, 55th ETRAN Conference, Jun 2011

57. Dj. Klisic, M. Zlatanovic,I. Radovanovic, *"WindSim Computational Flow Dynamics Model Testing Using Databases from Two Wind*

Measurement Stations", ELECTRONICS Vol. 15, Number 2, pp.43-48, Dec 2011

58. Comparative Study of Various Touchscreen Technologies, M.R. Bhalla, A.V. Bhalla, International Journal of Computer Applications (0975 – 8887) Volume 6– No.8, September 2010

59. http://gps.toptenreviews.com/navigation

60. http://www.ipcmax.com

61. Zoran R. Andjelkovic, Dragan R. Milivojevic, Zoran M. Stevic, Thermovisual camera commands decoding and ISI format encrypting, Journal of Scientific and Industrial Research, Vol. 69, (2010)

62. www.infraredsolutions.com

63. Zoran Stevic, Dubravka Nikolovski, Branislava Matic, Computerized Infrared Thermography in Distance Monitoring of Aging People, Proceedings of the XI International scientific-practical conference "Modern information and electronic technologies", Odessa, Ukraine, 24-28 May 2010, p 189

64. Siddiqi, U.F.; Shiraishi, Y.; Sait, S.M.;, "Multi-constrained route optimization for Electric Vehicles (EVs) using Particle Swarm Optimization (PSO)," *Intelligent Systems Design and Applications (ISDA), 2011 11th International Conference on*, vol., no., pp.391-396, 22-24 Nov. 2011

65. Siddiqi, U.F.; Shiraishi, Y.; Sait, S.M.;, "Multi constrained Route Optimization for Electric Vehicles using SimE," *Soft Computing and Pattern Recognition (SoCPaR), 2011 International Conference of*, vol., no., pp.376-383, 14-16 Oct. 2011

Chapter 2

Present and Future Role of Battery Electrical Vehicles in Private and Public Urban Transport

Adolfo Perujo[1], Geert Van Grootveld[1] and Harald Scholz[1]

[1] European Commission, Joint Research Centre, Institute for Energy and Transport, Sustainable Transport Unit, Ispra (Va), Italy

1. INTRODUCTION

"Electricity is the thing. There are no whirring and grinding gears with their numerous levers to confuse. There is not that almost terrifying uncertain throb and whirr of the powerful combustion engine. There is no water circulating system to get out of order – no dangerous and evil-smelling gasoline and no noise."[1] -

The OECD estimates that more than 70% of the developed world population lives in urban environments [2] - , which explains a larger concentration of vehicles there. In the EU-27, there were about 230 million passenger vehicles in 2007 and the new vehicle sales were nearly 16 million vehicles in that year. Notwithstanding the improvements in regulated air pollutants from road transport, the urban population remains at higher risk levels by directly suffering the impact of conventional vehicles because of their closeness to the pollutant source. On one hand urbanization means that people when travelling in their urban environment will typically travel less than 100 km a day. And on the other, that a large percentage of all transport and delivery of goods will take place in urban areas. Acceleration and deceleration frequency, traffic jams, thus energy efficiency and pollution per km are worst within urban traffic. Many business cases exist for urban electrified road transport because these offer a lower Total Cost of Ownership (TCO) than conventional means already today. The above reasons make the urban area the cradle where the electrification of road transport can deploy its full potential of positive impact, both environmentally and energetically.

There are several bottlenecks on the take-up by economic operators and the public at large of this technology, mainly: price of purchasing of an electric vehicle (EV), its limited range (range anxiety) and long charging time. Most of them are related to the present available battery technology. Improved batteries, maybe together with super-capacitors (so called hybrid power-packs) will most likely represent the core of the developments. The integration of the electrically recharging vehicle into the smart electric grid of the future, which calls for automatic communication technologies, is another frontline of research. Advances in these areas will probably reduce the obstacles for battery powered EVs in near future.

In the last 30 years the batteries' energy density (Wh/kg) has increased by a factor of four in three very well distinctive development waves: i.e. the development in 1995 of Ni-Cd batteries (with about 70 Wh/kg), that of Ni-MH in 2000 (~100 Wh/kg) and the third wave with the development of Li-ion batteries in 2005 leading to currently about 200 Wh/kg. With the present battery’s energy density a pure battery electric vehicle (BEV) can drive ca. 150 km with one charge, already opening the door for a substantial portion of series-produced EV models notably in urban environments. This already achievable all-electric range is larger than most of the daily average distance of city dwellers (in the USA about 90% of automobiles travel about 110 km daily and in Europe this distance is even smaller, as GPS-coupled monitoring analyses of ten-thousands of urban based cars have meanwhile proven also experimentally).

In any introduction of a new technology the role of stakeholders (public, commercial and private) is very important and their needs have to be understood and addressed. Because of the role of EVs in reducing the level of ambient pollution in urban conglomerations, this chapter will also look to different efforts and programs that some stakeholders as for instance different municipalities and regional and national governments, are setting up in order to actively support and stimulate the introduction of EVs.

Finally, the chapter will address how the above developments will support the introduction of EVs in the urban environment; it will also describe how reduced TCO will translate into more business cases and how this will impinge in a more general electrification of public transport with the consequent improvement of urban ambient air quality, noise levels, etc.

2. CITY VEHICLES

There is a very noticeable development effort on small city vehicles indicating that for the automobile industry (OEMs) the urban area represents the main niche in order to roll out the electrification of road transport. This effort is a globalised one with examples not only in Europe, but also in the US, China, Japan and India. In many cases demonstration is implemented by consortia of OEMs, or OEMs together with a university, or in public-private partnership.

Table 1 gives some examples of these cars besides the already launched ones in the market like, e.g. in Europe, the Smart for two Electric Drive, i-Miev, Peugeot-ION, Citroen C-Zero, Think City, etc..

We can conclude that OEM's are focusing on specific market segments within cities:

- The Smart for two for instance is part of a Car sharing project in Amsterdam;
- The HIRIKO will be used in Bilbao (Spain) to study the interest of the public for 'mobility on demand';
- The Renault Twizy is focused on very low purchase price and young customers;
- The VW Nils and the Audi concept are focused on individual transport.

It is noteworthy that for the city cars the OEMs are in particular concentrating in pure electric vehicles (BEV). Also hybridization of small cars is in development, and some technologies involved in hybridizing down-sized conventional engines, like capacitor banks of a few hundred Farad of capacity, might be cross-fertilizing the advent of advanced technologies also for pure electric solutions.

In the appendix further information on market share, number of BEVs per country and other data is presented.

3. RECHARGEABLE ENERGY STORAGE SYSTEMS (RESS) FOR VEHICLES

Rechargeable Energy Storage Systems (RESS) in vehicles include a variety of technologies, each one providing different sizes and different levels of maturity/development. Among these technologies we can name: Electrochemical Storage (Batteries, capacitors and notably super capacitors), Fuel-cell (often containing also a buffer battery) electricity provision with e.g., a hydrogen or *on-board reformer*fuel storage system, and (more in a niche situation) Compressed Air Energy Storage (CAES), and Flywheels. It is noteworthy to indicate that whatever is the chosen RESS for *electrified* vehicles it will be a key enabling technology for the penetration of this class of vehicles, because it influences in a decisive way their weight, energy efficiency, maintenance complexity and thus longevity and usability – and thus generally their acceptance-level achievable in the market.

In figure 1 some RESS are presented. From this figure the benefits of a hybrid power pack can be seen. These packs combine a high power density of fuel cells and batteries with a high energy density of (super) capacitors. Also the flywheels can be located in this figure.

This section intends to give an overview on battery, super-capacitors and hybrid power-pack (batteries plus super capacitors) developments that in a near future will probably reduce the obstacles, questions and doubts that potential users might have, and thus helping to bridge the gap between early adopters of the technology and the public at large.

3.1. Batteries

There are many possible chemistries (battery technologies) that are considered as possible viable options to be used in an electrified vehicle (either BEV or HEV). They range from the very well-established, but comparatively heavy lead-acid batteries to others still in its research stage as Li-air, Al-air or Fe-air batteries passing through Li-ion batteries that represent currently the most used battery-type in commercial BEV.

Table 1. Some examples of small city vehicles either in the process of being launched into the market or at concept stage

Model	Characteristics	View
Peugeot BB	Concept car, 4 seats, range of 120 km	
VW E-Up	4 seats and a range of 130 km (announced to come on the market in 2013)	
Toyota IQ FT-EV	Range 150 km (it will be in 2012 on the market)	
Gordon Murray T-27	Range up to 160 km, weight under 680 kg, now entering the investment phase	
Kia Pop	Range of 160 km, still a concept car	
HIRIKO	New concept of urban mobility, developed by MIT, it will be introduced in Bilbao in 2013	
VW Nils	One seater, light weight city car. It is a concept, for 2020	
Audi City Car	It is still a concept car	
Mahindra REVAI	Range of 80 km and a lead battery. It is a cheap car, coming soon to the EU market.	
Visio.M city EV (BMW & Daimler)	The aim is to develop a car with low price and low weight	
Renault Twizy	A low priced and low weight (500kg) city car. The battery is leased. The range is 100 km. The car is on the market since 2012.	

It is not the intention of this chapter to give an exhaustive insight[4] - on the chemistries of each of these batteries but rather to indicate the advantages and drawbacks as well as the possible gains in the future of new battery types still at the laboratory stage in terms of cost and specific energy/power, as these will strongly influence the viability of electrified vehicles.

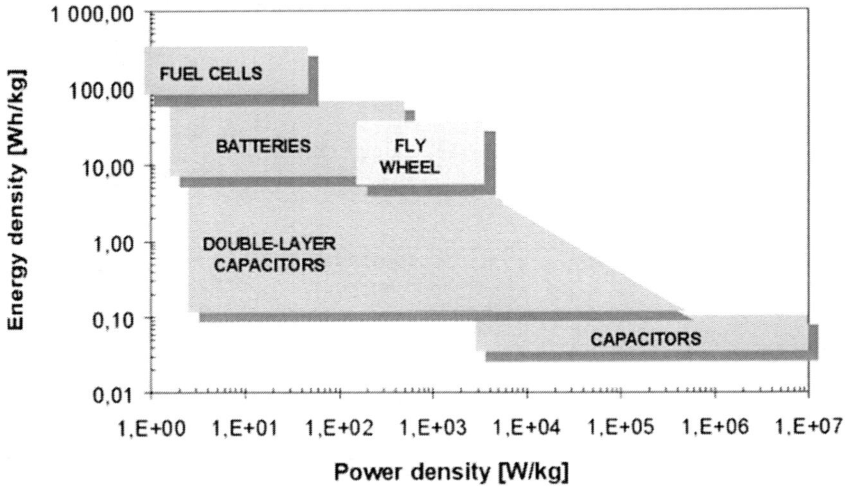

Figure 1. Energy density versus power density of different systems [3]

Figure 2 shows a possible battery technology development roadmap indicating some characteristics of the here discussed batteries technologies.

McKinsey argued in a paper that there are three important factors that could accelerate the development of electric vehicles. These are the manufacturing at (large industrial) scale, lower component prices, and boosting of battery capacity [1].

Table 2 shows some target performance parameters stated for batteries in electrified vehicles for the years 2015, 2020 and 2030 in a Technology Roadmap published by the IEA in 2009 [2].

Table 2 indicates that the energy density is expected to improve by a factor 5 and that the costs are expected to be reduced by a factor 10 within the next 20 years. These two parameters (energy density and costs) are seen to be the limiting factors of today's BEV. By increasing the energy density the range an electric vehicle can drive will be extended substantially leading to fewer stops for recharging. This should boost EV usability especially in typical urban use. Decreasing the costs of the battery will lead to substantially cheaper electric vehicles, enabling more purchases by the public and fleet investors, due to more sound business cases for commercial use of BEVs.

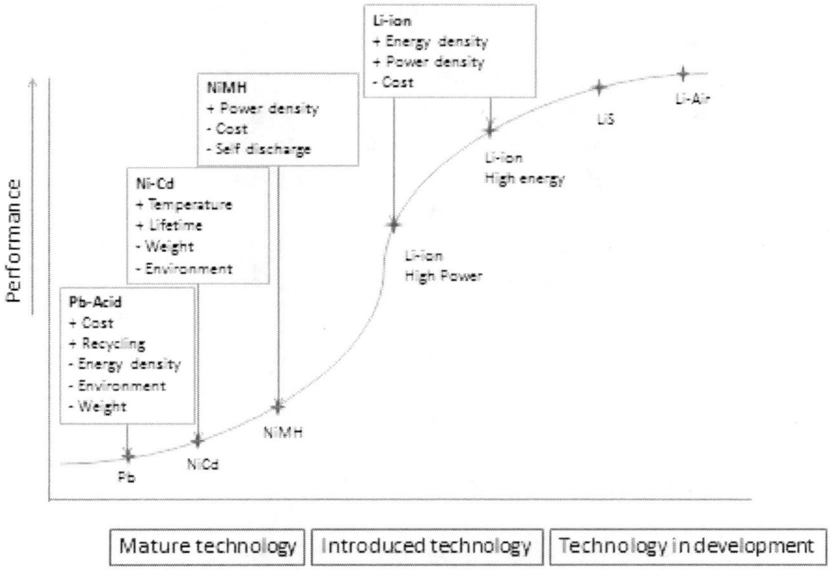

Figure 2. Battery technologies roadmap and characteristics

Table 2. Expectations on battery performances [2]

	Energy density (Wh/kg)	Power density (W/kg)	Costs (Euro/kWh)
2010	100	1000 - 1500	1000 - 2000
2015	150	1000 - 1500	250 - 300
2020	200 - 250	1000 – 1500	150 - 200
2030	500	1000 - 1500	100

The cycle-stability is an equally important parameter in applied battery chemistry. The attractiveness for automotive applications is not only dependent on the costs, the power density and the energy density of a battery, but also on the number of battery cycles that can be guaranteed.

3.1.1. Lead-Acid Batteries

The use of lead-acid batteries in electrified vehicles is mainly in industrial vehicles (e.g. forklifts, which must be heavy) because although at very affordable cost levels (100 – 150 $/kWh), the weight of lead representing about 60% of the weight of the battery translates into a low specific energy (30-50 Wh/kg), making this technology not competitive for most of electric road transport vehicles (even HEVs). It also suffers from a limited lifetime (3 – 5 years). It remains to be seen if lead-acid battery companies can substantially enter the market of micro-hybrid cars in view of small intermediate storage batteries as compared to the concurring battery technologies or modern, compact and lighter capacitor banks / supercapacitor units. At stake is a potential for growth of micro-hybridisation for small cars in the medium term (5-10 years).

3.1.2. Nickel-Metal Hydride Batteries

The use of Nickel-metal hydride batteries (NiMH) had been considered a sufficiently good intermediate stage for application in electrified vehicles (see e.g., the more than one million Toyota Prius sold with NiMH technology, and ca. two million hybrid cars running on NiMH world-wide.) Clearly outperforming NiCd batteries, they were the choice as long as there were still concerns on the maturity, safety and cost of Li-ion batteries. As NiMHs' specific energy (< 100 Wh/kg) cannot meet the requirements for full electric vehicles, it has been mainly used in hybrid vehicles (both HEVs and PHEVs) of limited storage capacity requirements. For PHEVs, NiMH on-board storage capacity arrived at electrical ranges of typically 30 km. There exist concerns on the supply of rare earths (typically mischmetal) and nickel in their anode respectively, cathode. The relatively high content of Ni and possibly rising Ni prices limit further the prospects of reducing their cost and thus use in future EVs.

3.1.3. Lithium-Ion Batteries

These batteries represent the most actual, wide-spread application in new BEVs world-wide. Nowadays BEVs with ranges above 150 km have all in common that the on-board storage is provided by Li-ion battery packs, often containing some sort of thermal control devices. The name of Li-ion batteries covers a large number of chemistries; indeed, if only a small number of them are actually in use, the list of potential electrode materials is quite large. On the other hand, possible electrolytes range from the mostly used solutions of lithium salts in organic liquids to ionic conducting polymers or ceramics additions to polymers. The current advantage position for this technology is based on its relatively high specific energy (it has reached 160 Wh/kg respectively, 450 Wh/l) however, at present, cost is still a drawback (700 – 1000 $/kWh). The main efforts are thus directed to decrease its cost and to increase its performance level keeping the system safe. There seems to exist a trade-off between performance of the cathode material and its safety. While cathodes made of $LiFePO_4$ depict good safety records its performance in terms of specific energy is poorer than, for example, $LiCoO_2$. However, the latter has a worse safety performance. $LiFePO_4$ also have a comparatively high amount of useful charging cycles during their life-time.

Present research concentrates on the development of an advanced Li-ion batteries exploring the capacity limits of the system through the development of new cathode and anode materials in combination with higher voltage (up to 5V) which will require new electrolytes and binders. Breakthroughs are expected from the combination of so called 5V or high capacity (and then lower voltage) new positive electrode materials and intermetallic new anodes [3].

3.1.4. High Temperature Na - B Alumina Batteries (Na-S And Na-$Nicl_2$)

The first prototypes of this battery type were introduced at the end of the 60s and contained sulphur as the positive electrode and the sodium β"-alumina as solid electrolyte. This material is an electronic insulator and exhibits sodium ion conductivity comparable to that of many aqueous electrolytes. However, to achieve enough electrochemical activity the Na-S battery operates between 300 and 350°C. Because of safety concerns, a derivative of this technology, based on the use of $NiCl_2$ instead of sulphur and termed ZEBRA battery [4], was later developed and evaluated for use in automotive applications. It has the advantage of being assembled in the discharged state and hence without the need of handling liquid sodium. As far as performance is concerned, its specific energy is relatively close to that of Li-ion batteries (115 Wh/kg), it has strongly improved its specific power (400 W/kg) and it has a relative low cost (600 $/kWh) although still between 4 to 6 time higher than the target set in many EV developing programs [5].

3.1.5. Other Battery Technologies

There are other battery technologies in the research stage that might in future meet the targets needed in electrification of road transport. We can mention among others Li-S [6], [7] and Li-air [8], [9] batteries (see figure 3). In particular they have demonstrated a specific energy[5] - about 300 Wh/kg. However, other aspects as life-time, achievable cycles over lifetime and specific power still need further research to meet the challenge.

In the line of using ambient air (oxygen) as the cathode, other materials such as Zn, Al and Fe can be used instead of Li. However, those systems are still in their infancy and at different stages of development. Developments on their recharge ability, air electrodes (porous design) cycle stability and safety are among the areas to be addressed.

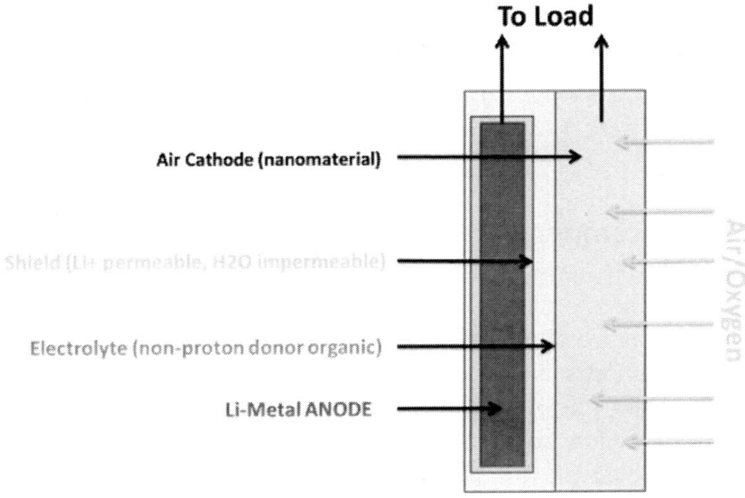

Figure 3. Scheme of a Li-air battery

3.2. Electrochemical Capacitors

These devices are sometimes referred to as 'ultra-capacitors' or 'supercapacitors' but these latter are rather commercial names.

Electrochemical Double Layer Capacitors refer to devices that store electrical energy in the electric double layer (EDL), which is formed at the interface between an electron conducting surface and an electrolyte. The EDL may be considered as a capacitor with two electrodes; the capacitance is proportional to the area of the plates and is inversely proportional to the distance between them. Their capacitance is very large because the distance between the plates is very small (several angstroms). The energy stored by such capacitors may reach 5 Wh/kg but they are power systems which can deliver their storage energy in a few seconds (up to 5s). Therefore, they are intermediates between batteries (high energy, low power density) and conventional capacitors (high power low energy density) and thus, they are complementary to batteries and are not in competition with them.

Supercapacitors are already used in transportation applications. They have been announced to be used in the starter/alternator of micro-hybrid cars and are under study by many car manufacturers (Toyota, BMW, Renault, PSA). Recently Ford and Ricardo UK announced[6] - the results of the HyBoost project, powering a small additional electric turbo-charging turbine for a down-sized thee-cylinder engine via such a fast ultracapacitor device of ca. 200 F capacity. Together with their outstanding cycle life, another key feature of EDLC systems is that, unlike Li-ion batteries, they can be recharged as fast as discharged. This is why they are used today in large-size applications for energy recovery in trams in Madrid, Paris, Mannheim and Cologne. There is hope, that a certain cross-fertilization in this area will happen between different improved road transport technologies, which may enable mass-production of EDLC systems sooner than later.

Supercapacitors (ultra-capacitors) have the ability to charge in a very short time however, its energy density is quite low and therefore by using only supercapacitors the electric range of an EV would not be sufficient. Consequently, the ideal situation would be combining both batteries and supercapacitors, which however requires a much more complicated voltage management.

3.3. Challenges

The performance of BEV and its competitiveness are closely linked to the performance of available battery systems in term of their specific power, efficiency and battery cost. In a recent paper Gerseen-Gondelach and Faaij [10] explored the performance of batteries for electric vehicles in the short and longer term. They review the different battery systems in term of performance and cost projection including sustainability aspects and learning curves. They concluded that well-to-wheel (WtW) energy consumption and emissions of BEVs are lowest for those with lithium-ion batteries, and that in the medium term only Li-ion batteries will have a specific power level of 400 W/kg or higher. Other battery systems like Li-S, Li-Air need efficiency improvements towards 90% to reach Well-to-Wheel (WtW) energy consumption of the BEV as low as found with Li-ion batteries. The author argued that already today, despite improvable efficiency levels, all batteries-types can enable similar or lower WtW energy consumption of BEVs compared to traditional internal combustion engine (ICE) vehicles: The WtW emissions are 20 – 55% lower using the EU electricity generation mix. Battery prices turned out to be of course the main parameter for improving the economics of BEVs e.g., if ZEBRA batteries attain a very low cost of 100 $ /kWh, such BEVs become cost competitive to diesel cars for driving ranges below 200 km. Such cost assumptions however were judged "unlikely" for the next and medium term.

With years of market introduction passing, an issue becoming provable will become battery ageing. With their use in extended time, batteries' performance can significantly reduce in terms of peak power capability, energy density and safety. Different auto manufacturers have set goals or targets for calendar life, deep cycle life, shallow cycle life and operating temperature range. However, it is still an issue of technological research to what extend current battery technologies can meet them.

Some examples of these targets are: for calendar life, the goals are typically for 15 years at a temperature of 35 °C, but current targets are for 10 years at which point a battery retains at least 80 per cent of its power and energy density. For deep cycle life, where the charge cycles go from 90 to 10 per cent of SOC[7] - , the goal is typically 5000 cycles, while the shallow cycle life expectation is 200,000 to 300,000 cycles. Goals for the temperature range as extreme as -40 to +66 °C can be found, such the question arises, whether batteries shall be specified for ambient conditions harsher than it has been done for any normal conventional ICE-vehicle. One extra difficulty that some of the results obtained on batteries performance are valid only for some specific charging and discharging rate and some specific range of ambient temperature exposure. It is still not clear if the test rates are more or less severe than the actual cycles a battery will be subjected to in an EV, and the interaction of ambient temperature with deep SOC cycling is also an unknown factor. A lot of pre-normative research is in front of us.

4. CITIES ARE THE NATURAL ENVIRONMENT TO DEVELOP AND TO IMPLEMENT E-MOBILITY

Cities are very important for the development and implementation of e-mobility, because the energetic and environmental benefits of BEVs replacing conventional vehicles are largest in city traffic. Moreover,

- About 70% of Europeans live in urban areas [11]. Most of the people live in cities with more than 50,000 inhabitants, and there are about 1,000 of such cities in Europe.

- Cities contribute substantially to the economics of Europe, 85% of European GDP is generated in cities [12].

- They contribute substantially to new knowledge (for instance from research being done on universities) and innovations by (high-tech) small and medium enterprises. Therefore cities have the potential to contribute to a better international competiveness of Europe.

The service sector is the most important source of employment in European urban economies. For example, in London, Paris, Berlin, Madrid and Rome the service sector accounts for between 80% and 90% of total employment. Examples of services are: government, telecommunication, healthcare/hospitals, waste disposal, education, insurance, financial services, legal services, consulting, information technology, news medias, tourism, and retail sales.

Providing and using these services lead to large transportation needs and activities of people and goods, and this, in turn, leads to a high use of energy and to the generation of anthropogenic emissions, like CO_2, NO_x, ozone, fine particles, noise, etc.

Let's focus in some of these aspects.

The energy consumption in European cities is high. About 80% of Europe's energy is used in cities [13]. It is expected [14] that this number will increase in future, because the urban population will grow and also the economic activities and the prosperity will grow.

We have about the same figures for CO_2. Cities are the largest emitters of CO_2. About 75% of the European's CO_2 is emitted in cities. On average the CO_2-emissions for European cities are in the neighbourhood of 1 ton CO_2 per capita per year [15]. Of course these emissions are dependent on the modalities of transportation which are used in the different cities. The higher the share in public transport, walking, cycling the lower the CO_2-emissions will be per capita.

Some examples: In Berlin [16], in 2008 32% of the people choose a car for transportation, 29% walked, 26% public transport and 13% took the bike. In London [17], in 2007 41% choose the car, 25% public transport, and 30% walked.

In some situations the concentration of NO_x and fine particles exceed the air quality limits. These situations are also called: hot spots. NO_x contributes to the formation of smog. Also acid rain can be formed out of NO_x.

Figure 4, depicts a street canyon in Copenhagen [18]. In many European cities the dispersion of air pollution is restricted by the geometry of buildings. This creates so-called street canyons. These canyons lead to elevated concentrations of local pollution, and therefore people living in (or in the neighbourhood of) these hot spots have a higher risk for getting ill.

Figure 5 depicts the concentration of NO_x in ambient air in a city (London) [19]. As can be seen from this picture the NO_x-concentrations exceed the maximum regulated value, which is 40 $\mu g/m^3$.

Figure 6 shows the NO_x-concentrations in European regions [19]. The intensively populated zones can be recognized easily. These are mainly cities and intensively used highways between the cities.

Figure 4. An example of a street canyon in Copenhagen [18]

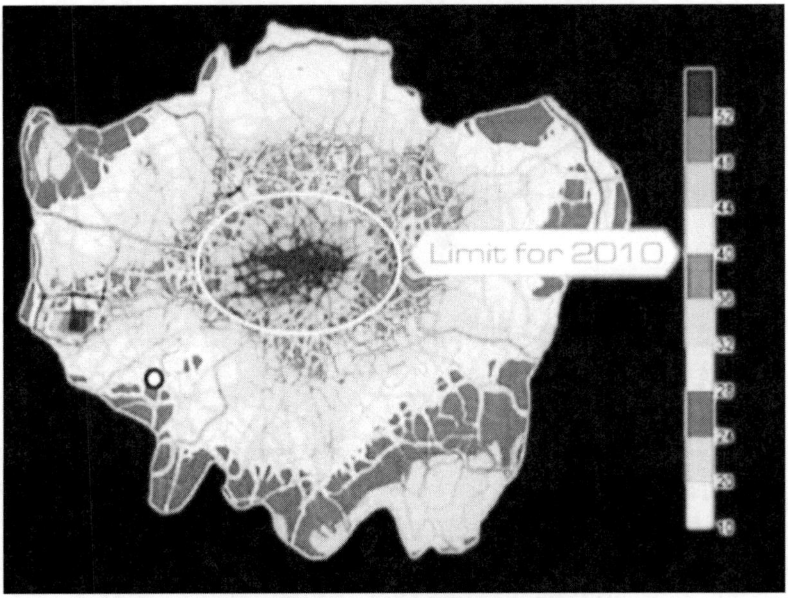

Figure 5. NO_x-emissions in the city of London [19]

4.1. E-Mobility Can Tackle These Problems; Many Stakeholders Are Willing To Contribute

The big advantage of e-mobility is that it gives direct results for improving ambient air quality. An electric vehicle does neither emit NO_x and PM, nor VOC (volatile organic compounds). So, when electric vehicles are introduced to replace conventional vehicles, these emissions decrease directly and ambient air quality will improve. Because ozone is formed by a photo-catalytic reaction between VOC and NO_x, also the ozone concentration will be reduced.

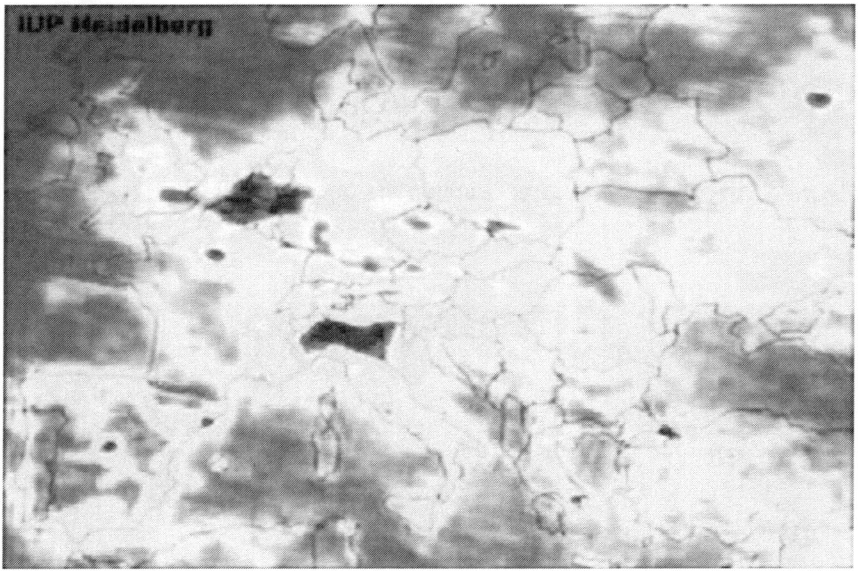

Figure 6. NO_x-emissions in Europe

A large number of stakeholders have parallel interests in the development and implementation of e-mobility in cities. The citizens want a clean city to live in. So, the ambient air quality needs to improve in several situations.

These ambient air problems are also a main driver for the politicians and administrations of cities to stimulate electro-mobility. The Covenant of Mayors which is signed on February 2009 is a good example of this. The main goal of this covenant, which now has about 4,000 signatories, is to increase energy efficiency and to use renewable energy sources. Within the framework of this covenant the Sustainable Energy Action Plans (SEAP) play a central role. A Sustainable Energy Action Plan (SEAP) is the key document in which the Covenant signatory outlines how it intends to reach its CO_2reduction target by 2020. Already more than 1400 of these plans are submitted. A lot of these plans contain actions on the stimulation of electro-mobility in cities.

Also business leaders are major stakeholders. A first reason for that is that e-mobility can lead to sound Total Costs of Ownership (TCO). This means that economic activities can be done more cost effectively with e-mobility than with the petrol based vehicles. A second reason is that the spin-off of this technological development can be enormous. It is already stated that there are about 1,000 middle large cities in Europe. This is a big market for small and medium sized enterprises that develop new technologies for implementing e-mobility-systems.

4.2. Cities As Living Labs: Some European Experiences

Cities can be regarded as a living lab. This means that they have the possibility to test new concepts under real life circumstances. The behaviour of consumers working with new concepts can be studied, and the feedback of the consumer can be used by the supplier to modify and improve the concept. So, a cyclic process can be organized leading to the rapid development of new concepts. The administrations can take the lead in organizing these processes. They have all the ingredients to do so: the consumers, the suppliers, the infrastructure, and also the challenges and the solutions.

There are a lot of interesting projects going on in European cities on the development and implementation of electric vehicles. Some examples are the projects started within the European Green Cars Initiative [22].

Most of them concern electric mobility, for instance the Green eMotion project [23]. This project is supported by 43 partners from industry, the energy sector, electric vehicles manufacturers, municipalities as well as universities and research institutions. The goals of Green eMotion are:

- Connecting ongoing regional and national electro mobility initiatives;

- Comparing the different technology approaches to ensure the best solutions prevail for the European market;
- Creating a virtual marketplace to enable the different actors to interact;
- To demonstrate the integration of electro mobility into electrical networks (smart grids);
- Contribute to the improvement and development of new and existing standards for electro mobility interfaces.

In several projects ICT is introduced to facilitate the implementation of electromobility. One of these is the project MOBI.Europe [24]. In this project the users of electric vehicles are getting access to an interoperable charging infrastructure, independently from their energy utility and region. It is built on the e-mobility initiatives of Portugal, Ireland, the Spanish region of Galicia and the Dutch city of Amsterdam.

Another project is the smartCEM [24] project in which four European cities/regions are participating: Barcelona (ES), Gipuzkoa-San Sebastian (ES), Newcastle (UK) and Turin (IT). The goal of this project is to demonstrate the role of ICT [8] - solutions in addressing shortcomings of e-mobility, by applying advanced mobility services, like EV-navigation, and EV-efficient driving.

One part of the VIBRATe (VIenna BRATislava E-mobility) [25] project is to identify the possibilities of connecting two neighboring metropolitan areas—Bratislava (Slovakia) and Vienna (Austria) with a “green” highway. This highway will interconnect the two cities with a network of public charging stations for electric vehicles. In this project IBM is working together with Západoslovenská energetika, a.s. (ZSE) and the concerned municipalities.

Autolib [26] is an electric car-sharing program which is launched in Paris at the end of 2011. This program will start with 250 vehicles. The amount of vehicles will grow to 2,000 in the summer of 2012. This number will grow to 3,000 in the summer of 2013. In this car-sharing program the compact Blue car is introduced. This four-seat car is the result of a collaboration of the Italian car designer Pininfarina and the French conglomerate Groupe Bollore.

Car2Go [27] is a subsidiary of Daimler AG that provides car sharing services in several cities in Europe and North America. In November 2011 a fleet of 300 smart for two electric vehicles was deployed in Amsterdam.

In London the "Electric 10" is formed. This is an initiative of 10 companies that use electric commercial vehicles for their activities. The Electric 10 partnership was formed in autumn 2009, bringing together 10 major companies who are already using electric fleet vehicles on daily basis: Sainsbury's, Tesco's, Marks and Spencer, UPS, TNT Express, DHL, Amey, Go Ahead, Speedy, Royal Mail. The Municipality of London is working with these companies to learn from their experiences and encourage others to take their lead [28]. The use of electric vehicles for goods delivery not only benefits the environment, it also has a positive total cost of ownership (TCO).

4.3. Cities Have The Power To Implement; And They Are Already Doing So

City administrations have the possibility to develop new concepts under real life circumstances, as we have seen in paragraph 4.2. and to set projects to bring e-mobility to a reality. Of equal importance is that they also have the ability to implement using their legal instruments. Many cities are already doing this.

The instruments they use can be divided in three categories [29]:

Financial incentives

Examples of financial incentives are exemptions from vehicles registration taxes or license fees. Or exemptions from congestion charge. Another financial incentives are of operational nature e.g. the electric vehicle gets a discount on parking costs.

Non-financial incentives

There are cities which give non-financial incentives. For instance, free or discount cost for a parking place in the city centre. Or that the owner will get access to restricted highway lanes. An important incentive is also to get easy access to public charging facilities.

Their purchasing power

Municipalities are not only regulators. They also have a vehicle fleet and they give licenses to public transport systems. With these possibilities they also can stimulate the e-mobility. They can buy electric vehicles for their municipal fleet and they can add hybrid buses to public transport systems. Municipalities can install charging stations on the public area, like: libraries, parking garages, city halls, or other public buildings.

4.4. Some Remarks To This Section

In paragraph 4.2 a total of 7 projects which are presently going on in Europe are described shortly. It should be stated that these are just illustrations. There are many more interesting projects on e-mobility. What we see is a steep increase in the amount of battery electric vehicles (BEV) in Europe [30]. In 2010 in total 765 BEVs were introduced on the EU-27-market, and in 2011 already ca. 9,000 BEVs. This took place predominantly in France, Germany, UK, the Netherlands, and Austria. The main BEVs types were Peugeot-ION, Mitsubishi-i-MIEV, Smart for two, Nissan-Leaf, and Citroen-C-Zero.

We expect that this steep increase will continue, because of the battery developments we described in the beginning of this chapter and also because of the strong efforts of stakeholders, like member states, municipalities, car manufacturers, and the EU. Indeed, in the 2011 Transport White Paper 'Roadmap to a Single European Transport Area – Towards a competitive and resource efficient transport system' (COM (2011) 144 final), the European Commission proposes 10 goals for a competitive and resource efficient transport system which serve as benchmarks for achieving the 2050 60% GHG emission reduction target. One of these goals is to halve the use of 'conventionally-fuelled' cars in the urban transport sector by 2030 and to phase them out by 2050, thereby also reducing the transport system's dependence on oil.

5. ENABLING TECHNOLOGIES FOR THE INTRODUCTION OF ELECTRICITY IN ROAD TRANSPORT

Another reason why electric vehicles are promising is because of the fact that it can contribute to the development and introduction of smart grids. With smart grids the share of green electricity by means of wind and solar can be better managed to increase. Electric vehicles can serve as storage for electricity (spinning reserves) in those times when the households don't need the amount of electricity produced at a certain moment, and the vehicles can deliver electricity to the grid in times when the households need more electricity than produced at that moment. The benefits are that with these smart grids the CO2 emissions will decrease as well as the use of fossil energy. The CO2 emissions will go down even more, because from well-to-wheel-analyses it can be seen that in most cases the CO2-performance of electric vehicles is better that petrol based vehicles [21].

It is generally considered that smart grid and V2X where X represents another vehicle (V2V), the grid (V2G) or sometime the user's home (V2H) are essential technologies for the early introduction of electrified vehicles as these provide an added value to the vehicle respectively, reduce its TCO.

5.1. What Is A Smart Grid?

The concept of Smart Grid[9] - was developed in 2006 by the European Technology Platform for Smart Grids, and concerns an electricity network that can intelligently integrate the actions of all actors connected to it - generators, consumers and those that do both - in order to efficiently deliver sustainable, economic and secure electricity supplies. Interoperability of EVs to Smart grids promises an increase of the EVs' overall energy efficiency and cost benefits.

Decentralized supply of electricity is growing. There are at least three types of decentralized supply options:

- More and more wind turbines are in operation;
- (micro) Combined Heat and Power (CHP) is up-coming.
- The generation by means of solar PV is increasing;

This development means that the fluctuations over time in the supply of electricity would be increasing in a near future with the consequent challenge to harmonize it with the demand of electricity.

There are some options to deal with this challenge. The first one is to influence the regulate the supply. When at a certain moment more wind and solar electricity is produced then the supply of electricity from fossil sources should be limited. To realize this real-time communication between consumers and producers should take place. This can be done by means of smart meters. To reduce the supply of electricity from fossil sources is, however, not always an easy task.

A second option is to realize a situation in which the fluctuations which might appear on the supply side will be match on the demand side. This can be realized by introducing fluctuating prices, which again can be realized by means of smart meters. So when the supply is high then the price will be low and then the smart meter can for instance start charging an electric vehicle or it can start other appliances e.g. the washing machine. And when the demand is high then the price will be high and then the electric vehicle will supply electricity to the house. Thus, by means of the price mechanism and the smart metering the supply and the demand can stay in balance, despite of the fluctuations occurring in the supply side.

A third option is by introducing a storage facility. This can be done by means of fly-wheels, ultra-capacitors, compressed air, and batteries. In this third option the batteries of the electric vehicles can play a role (see section 3).

A fourth option is that when there is an oversupply of electricity, it is used for the electrolysis of water and the formed hydrogen is either used directly or it is coupled with CO2 to produce methane. When there is a shortage of electricity the hydrogen and/or methane can be used to produce electricity by means of a Combined Heat and Power Plant (CHP). Of course the hydrogen can also be used to fuel a Fuel Cell Electric Vehicles (FCEV's).

Hence, with a smart grid it is possible to:

- Better facilitate the connection and operation of electrical generators of all sizes and technologies;
- Allow consumers to play a part in optimizing the operation of the system;

- Provide consumers with more information and options for choosing an energy supply;
- Significantly reduce the environmental impact of the whole electricity supply chain;
- Organize a symbiotic relation between the grid and the electric vehicle. The vehicle can be charged when the price is low, and the vehicle can contribute to the grid when electricity is needed there;
- Charge the electric vehicle with low-CO2-containing electricity, which contributes to low CO2-emissions when driving the vehicle;
- Maintain or improve the existing high levels of system reliability, quality and security of supply;
- Maintain or improve the efficiency of existing services;
- Foster the development of an integrated European market.

5.2. Some European Efforts On A Practical Scale On Smart Grids Including Electric Vehicles

There are considerable efforts in Europe (the same thing can be said on other developed markets; i.e. USA, Japan...) on smart grids[10] - by supporting and carrying out many projects. In several of these projects electric vehicles are included and studied. Some examples are:

InovCity concept in Évora (Portugal)

The goal of this project is that the entire municipality of Évora will be connected to an intelligent electricity system which includes 30,000 customers.

Some characteristics of this project are:

- The project is initiated by EDP[11] - Distribuição, with support from national partners in industry, technology and research (EDP Inovação; Lógica; Inesc Porto; Efacec; Janz and Contar);

- The electricity grid is provided with ICT, so that the grid can be controlled automatically. This is done by monitoring the grid in real time;

- Increase of renewable energies (PV solar cells, micro wind turbines) is facilitated by the intelligent electricity grid;

- The Energy Box plays a central role in this system. All consumers will have such a box, and this box connects the consumers to the intelligent grid. In the box the amount of electricity used and/or produced is recorded. And by means of the box the consumers can program devices, like washing machines, when the price of electricity is low. This process of programming devices can be automated fully;

- The electricity grid is also facilitating the charging and discharging of electric vehicles. The batteries of these vehicles will serve as a buffer when there is an oversupply and the batteries will serve as a producer of electricity when more electricity is needed in the homes.

Endesa's Smartcity Málaga Project (Spain)

The goals of SmartCity Málaga[12] - are to implement and integrate distributed energy resources, energy storage, electric vehicle charging and discharging facilities, and intelligent public lighting devices.

The characteristics of the project are:

- Endesa in cooperation with 11 partners is rolling out state-of-the-art technologies in smart metering, communications and systems, network automation, generation and storage, and smart recharging infrastructure for e-vehicles;

- More than 17,000 smart meters are installed;

- 11 MW of renewable generation capacity which consist of solar PV, wind energy and cogeneration;

- A storage facility consisting of batteries;

- A network of recharging points for vehicle-to-grid-technology;

- By means of ICT all these devices are connected to the Network Control Center, where these are monitored and controlled.

Harz.EE-Mobility (Germany)**[13]**

This project has been initiated by Siemens CT in cooperation with 14 partners – including research institutes, the Deutsche Bahn (German Railroad Company), and wireless provider Vodafone. The goal is to make Germany's Harz district a model region for electric mobility. Wind, solar, and other alternative energy sources already contribute more than half of the power generated in the Harz district. Sometimes in windy periods some wind turbines have to switch off. This problem could be solved using electric vehicles as small energy storage units allowing for useful demand shift.

The project focuses on Vehicle-to-grid-technology (V2G). Electric cars would recharge their batteries whenever winds are strong, especially at night. Conversely, during calm periods they could feed electricity back into the grid at higher prices. Ultimately, V2G aims at bidirectionality of both, car / grid communication and their energy flow.

In this project an energy management system is developed. All the 2,000 energy generation devices are connected and automatically controlled (PV, wind turbines, biogas, and electric vehicles). The project also monitors and studies the movement profiles of electric vehicles. With this information is can be predicted how many electricity in what period is needed to recharge the vehicles. This will also be important control data for the electricity generation devices.

The above examples indicate the important role that information and communication technology play in an early uptake of electrical vehicles by their seamless integration in the electrical distribution and control network ("smart grid" of the future).

6. DISCUSSION

This chapter has focussed on the technological requirements that electrical vehicles need in order to break into the (primarily urban) main stream as a valid personal or commercial transport means. However, their cost/price and environmental impact have not been addressed. This section intends to indicate some of the recent efforts that can be found in the open literature, both to forecast when this vehicle technology will become possibly a preference of the user and what policies could be put in place to better address the environmental benefit of increasingly electrifying road transport.

In a recent paper [31] Weiss *et al.* have forecasted the price for hybrid-electric and battery-electric vehicles using *ex-post* learning rates for HEVs and *ex-ante* price forecasts for HEVs and BEVs. They forecasted that price breakeven with these vehicles may only be achieved by 2026 and 2032, when 50 and 80 million BEVs, respectively, are expected to have been produced worldwide. They estimated that BEVs may require until then global learning investments of 100–150 billion € which is less than the global subsidies for fossil fuel consumption paid in 2009. Their findings suggested that HEVs, including plug-in HEVs, could become the dominant vehicle technology in the next two decades, while BEVs may require long-term policy support. In line with what it has been pointed out in this chapter, the authors indicated that the performance/cost ratio of batteries is critical for the production costs of both HEVs and BEVs. If current developments persist, vehicles with smaller, and thus less costly, batteries such as plug-in HEVs and short-range BEVs for city driving could present the economically most viable options for the electrification of passenger road transport until 2020.

More studies on specifically urban electrification of road transport might move the quantitative arguments to some extent, and show that there are several niches of earlier cost-effectiveness even for BEVs.

There is a debate on how to consider the environmental impact of this class of vehicles. Unlike their counterpart fossil-fuelled vehicles, the emissions generated by electrified vehicles are produced “upstream”, that is where the electricity is generated. Should they be considered to have GHG emissions of "0 g/km"? Lutsey and Sperling [32] argue that considering electric vehicles as 0 g/km and assuming 10% of cars sold by 2020 to be electric, this could result in a loss of 20% of the conventionally calculated benefit from USA regulations aimed at reducing vehicle GHG emissions – so one has to pay attention of what is summed up. They also found that if upstream emissions were included, an electric vehicle powered from the America electricity grid would on average emit about 56% less CO2 than their petrol counterpart (104 g/mile compared to 238 g/mile). It is clear that the exact amount will depend upon the particular electricity generation fuel mix and thus generation efficiency in the given State where the BEV was charged. These authors support the idea of using a full lifecycle analysis as regulatory option rather than the "0 g/km". This approach, although more complicated, would ensure that GHG regulations were scientifically rigorous and could accommodate future energy technology development.

7. CONCLUSIONS

In the 2011 Transport White Paper 'Roadmap to a Single European Transport Area – Towards a competitive and resource efficient transport system' (COM (2011) 144 final), the European Commission proposes 10 goals for a competitive and resource efficient transport system which serve as benchmarks for achieving the 2050 60% GHG emission reduction target. One of these goals is to halve the use of 'conventionally-fuelled' cars in the urban transport sector by 2030 and to phase them out by 2050, thereby also reducing the transport system's dependence on oil. Among the possible options to support this target, the electrification of road transport seems to be a winning one - as we have indicated in this chapter. We have addressed the technological challenges that electrified vehicles have to face in order to overcome the present status quo. These are mainly due to the storage system on board of a BEV. There are promising technologies that can positively support the introduction of electric vehicles in our streets and roads (e.g. V2G and interoperability with smart grids through standardised communication). Finally, the areas of cost and environmental impact has been also addressed by commenting recent efforts in both forecasting the price reduction in the future and addressing full life-cycle analysis as possible policy options to include the full picture of the impact of vehicles in GHG emissions.

NOTES

1. Thomas Alva Edison (February 11, 1847 – October 18, 1931)

2. See e.g., p. 17 in " Trends in Urbanisation and Urban Policies in OECD Countries: What Lessons for China?", OECD and CDRF, http://www.oecd.org/urban/roundtable/45159707.pdf

3. http://www.mpoweruk.com/alternatives.htm

4. For a more exhaustive review of storage technologies see e.g.,: "Outlook of Energy Storage Technologies" (IP/A/ITRE/FWC/2006-087/Lot 4/C1/SC2) and "White Paper. Battery Energy Storage Solutions for Electro-mobility: An Analysis of Battery Systems and their Applications in Micro, Mild, Full, Plug-in HEVs and EVs" EUROBAT Automotive Battery Committee Report.

5. This value is at cell level (research object) as for a battery pack is expected to be lower due to the extra weight of materials used for packing and interconnection of the cells.

6. http://www.theengineer.co.uk/in-depth/analysis/hyboost-programme-promises-engine-efficiency/1010742.article.

7. SOC means State of Charge

8. ICT = Information and Communication Technologies

9. http://www.smartgrids.eu/documents/TRIPTICO%20SG.pdf

10. A survey can be found on http://www.smartgridsprojects.eu/map.html

11. http://www.inovcity.pt/en/Pages/media-center.aspx

12. http://www.endesa.com/en/aboutEndesa/businessLines/principalesproyectos/Paginas/Malaga_SmartCity.aspx

13. *http://www.siemens.com/corporate-technology/en/research-cooperations/mobility.htm*

REFERENCES

1. Russell Hensley, et al. Battery technology charges ahead. McKinsey Quarterly. July 2012

2. Technology Roadmap, Electric and plug-in hybrid electric vehicles, International Energy Agency, 2009 (page 43).

3. Developments in battery chemistries: 2009 European Energy Fair Zürich: www.investorrelations.umicore.com

4. P. Tixador; IEEE/CSC & ESAS EUROPEAN SUPERCONDUCTIVITY NEWS FORUM, No. 3, (2008), www.esas.org

5. Matsushita, EP 0569 037, 7 July 1993

6. Mikhaylik Y. V. et al. High Energy Rechargeable Li-S Cells for EV Application. Status, Challenges and Solutions. ECS Trans. 2009; 25 (35) 23-34

7. Kolosnitsyn V. S. and Karaseva E. V. Lithium-sulfur batteries: Problems and solutions. Russian Journal of Electrochemistry, 2008; 44 (5) 506-509

8. Abraham K. M. and. Jiang Z. A Polymer Electrolyte-Based Rechargeable Lithium/Oxygen Battery. J. Electrochem. Soc. 1996; 143 (1) 1-5

9. Advanced Batteries Technology, March 2010, including announcement of IBM project launch

10. Gerssen-Gondelach, Sarah J., Faaij, André P. C. Performance of batteries for electric vehicles on short and longer term. Journal of Power Sources 2012; 212 (0) 111-129

11. Promoting sustainable urban development in Europe ACHIEVEMENTS AND OPPORTUNITIES, http://ec.europa.eu/regional _policy/sources /docgener/presenta/urban2009/urban2009_en.pdf

12. EUROCITIES:http://www.eurocities.eu/eurocities/issues/climate-adaptation-issue (accessed 3 July 2010).

13. EUROCITIES: http://www.eurocities.eu/eurocities/issues/energy-efficiency -issue (accessed 3 July 2010).

14. International Energy Agency: World Energy Outlook 2008: Chapter 8. Energy use in cities http://www.iea.org/weo/docs/weo2008/ WEO_ 2008_Chapter_8.pdf (accessed 3 July 2010).

15. International Association of Public Transport: PUBLIC TRANSPORT AND CO2 EMISSIONS http://www.uitp.org/news/pics/pdf/MB_CO23.pdf (accessed 3 July 2010).

16. City of Berlin: Mobility in the City: Berlin Traffic in Figures: Edition 2010 http://www.stadtentwicklung.berlin.de/verkehr/politik_planung/zahlen_fakten/download/Mobility_en_komplett.pdf (accessed 3 July 2010).

17. Transport for London. Travel in London Key trends and developments. Report number 1. http://www.tfl.gov.uk/assets/downloads/corporate/Travel-in-London-report-1.pdf (accessed 3 July 2010).

18. ASSET (Assessing Sensitiveness to Transport): Sensitive urban areas http://www.asset-eu.org/doc/Copenhagen.htm (accessed 3 July 2010).

19. Photo courtesy of Wikipedia. The picture is derived from www. dft.gov.uk

20. REBOURS, Yann and KIRSCHEN, Daniel. What is spinning reserve? The University of Manchester report (Release 1, 19/09/2005). Available at: http://www.eee.manchester.ac.uk/research/groups/eeps/publications/reportstheses/aoe/rebours%20et%20al_tech%20rep_2005A.pdf.

21. Edwards, R., Larive, J.F., Beziat, C. Well-to-wheels Analysis of Future Automotive Fuels and Powertrains in the European Context. WELL-to-WHEELS Report Version 3c, July 2011. European Commission Joint Research Centre, Institute for Energy.

22. European Green Cars Initiative. http://www.green-cars-initiative. eu/about (accessed 3 July 2010).

23. Green eMotion. http://www.greenemotion-project.eu/about-us/index.php (accessed 3 July 2010).

24. ElectroMobility pilot projects launched at the European Parliament (February 2012)http://www.icarsupport.eu/media/news-2/electro-mobility-pilot-projects-launched-at-the-european-parliament/(accessed 3 July 2010).

25. Green Car Congress. Energy Technology Issues and Policies for Sustainable Mobility. http://www.greencarcongress.com/2012/04/vibrate-20120409.html (accessed 3 July 2010).

26. European Travel. http://www.msnbc.msn.com/id/45553670/ns/travel-destination_travel/t/paris-launch-electric-car-sharing-program/ (accessed 3 July 2010).

27. Wikipedia, the free encyclopaedia. http://en.wikipedia. org/wiki/ Car2Go (accessed 3 July 2010).

28. NYC Global Partners' Innovation Exchange website. Best Practice: Electrical Vehicle Development http://www.nyc.gov/html/unccp/ gprb/ downloads/pdf/London_ElectricVehicles.pdf (accessed 3 July 2010).

29. Forbes. The Global Electric Vehicle Movement: Best Practices From 16 Cities. http://www.forbes.com/sites/justingerdes/2012/05/11/the-global-electric-vehicle-movement-best-practices-from-16-cities/ (accessed 3 July 2010).

30. E-Mobility in the EU, Facts and Figures, European Commission, DG Joint Research Center, to be published

31. M. Weiss, et al. On the electrification of road transport - Learning rates and price forecasts for hybrid-electric and battery-electric vehicles. Energy Policy (article in press)

32. Lutsey N. and Sperling D. Regulatory adaptation: Accommodating electric vehicles in a petroleum world. Energy Policy 2012; 42 308-316

Chapter 3

Is the Electric Vehicle an Attractive Option for Customers?

Israel García * and Luis Javier Miguel

Systems Engineering and Automatic Control Department, University of Valladolid, Paseo del Cauce s/n. 47011 Valladolid, Spain

ABSTRACT

As a new technology, electric mobility has the potential to achieve a reduction in CO_2 emissions and contribute to the transition from the current transportation system to a better one, environmentally speaking. The objective of the paper is to aid the necessary decision-making for the adoption and development of electric vehicles in Spain, taking the time horizon of 2020. This will be achieved by building a System Dynamics model for various scenarios that will be used for the analysis and comparison of various dynamic variables, as well as to determine how, and to what extent, they will influence the number of electric vehicles that will run on Spanish roads in the coming years, focusing on the cost variable.

KEYWORDS

electric vehicles; batteries; CO_2 emissions; electric vehicle costs

1. INTRODUCTION

Road transport today is responsible for a significant and growing share of global CO_2 emissions. In 2008, the transport sector consumed 38% of the total energy used in Spain. Within the sector road transport is still almost entirely dependent on oil-derived fuels (about 98%), and is therefore highly vulnerable to possible oil price fluctuations and supply disruptions. The transport sector is responsible for more than 25% of CO_2 emissions in Spain, and road transport is responsible for 90% of the CO_2 emissions in this sector, with an energy consumption of about 80% of the sector's total [1].

Looking at these data, a change process in the automotive sector which involves a reduction in both CO_2 emissions and energy consumption of vehicles running in Spain is necessary. To achieve this, the government has begun to develop strategies to drive the change, among which one is the promotion of alternative propulsion technologies to conventional ones (MOVELE project).

A transition to EV (electrically-powered vehicles) has been proposed because they have the potential for significantly reducing CO_2 emissions. This can be justified, in the first place, since the electric motor is more energy efficient than the internal combustion engine [2]. The "Well-To-Wheel" model is used to compare emissions from both engines [3]. Besides the direct CO_2 emissions from the exhaust, the use of a vehicle also causes indirect CO_2 emissions emanating from the fuel supply chain. The sum of the direct emissions and the emissions from the fuel chain are the Well-To-Wheel. In the case of the internal combustion engine, only a performance that at best does not reach 20% can be obtained, while for the electric vehicle, with electric power from the current mix, it is about 30% [2,4]. The electric vehicle produces no polluting emissions during its use due to the electric motor. However, electric vehicles will not be completely exempt from emissions as such. There will be some emissions to compute, depending on where electrical energy is obtained. Accordingly, for the electric vehicle to generate no emissions during its use, the electricity must come from CO_2-free sources, such as renewable sources.

As expected, the change will not be easy. The task is certainly complex: changing consumer habits after many years of using one kind of vehicle with a range of benefits that, in some way, will cease to exist.

Regarding the conventional vehicle, electric vehicles currently available usually have a higher purchase price and shorter lifetime than conventional power trains [5]. Electric vehicles are in the early phases of the adoption process [6] and it is important to analyze whether the characteristics of the electric vehicle actually meet the consumers' needs.

As mentioned, the first deterrent the customer finds is the purchase price difference (essentially the cost of the battery) described by previous works [7]. However, this cost is compensated by the lower cost of electricity compared to gasoline and the lower cost of maintenance [8]. This paper examines individuals' motivations when purchasing vehicles, focusing upon what economic factors would encourage consumers to purchase an electrical vehicle or a conventional vehicle.

This research attempts to ascertain, from an economic point of view, if reductions in fuel costs, mechanical damage to the engine (due to its simplicity), or greenhouse gas emissions would encourage individuals to purchase an electric or a conventional vehicle in the next 10 years. To do this, we consider no only the costs when acquiring the vehicle, but also the costs over its useful life (including oil and electricity price fluctuations, taxes, garage conditioning for recharging, *etc*.), in order to determine the years of repayment on the purchase of an electric vehicle compared to a conventional one.

Nowadays, there is much literature on electric vehicles, including some very detailed works, such as "*On the Road 2035*" [2], the Electric Power Research Institute (EPRI) Reports [9,10,11], and the research by the Institute of Transportation Studies at UC Davis [12,13]. This paper, however, mainly examines the economic influence at the time of purchasing the vehicle, but also the costs over its lifetime use, focused on the Spanish context. Regarding the Spanish sector, there are not many policymaker studies [14], and even fewer focused on the policies necessary for the integration of electric vehicles in society and the methodology used, a question that motivates the creation of this paper.

2. WORK METHODOLOGY

The potential future developments of the electric vehicle sector are still unknown, making it impossible to analyze all the influential elements, so to forecast the adoption process of the electric vehicle, an approximation must be used. In a previous work by Geels [15], it is explained that the conceptual perspective of the technology transition to the electric vehicle is fairly complex. "Can it be made operational for empirical research? The proof of the pudding is in the eating, *i.e.*, use the perspective for empirical analyses of dynamics of socio-technical systems". This means that there is an abstract representation of technology transitions. However, a more empirical version of this process is required.

The methodology used in this project is the creation of a simulation model through which different policy options are played out in different scenarios, leading to different results. Due to the complexity of the model and the variety of interactions between its component variables, the research method used to create the model and its analysis is System Dynamics. System Dynamics is a method that challenges us all how to move from generalizations (abstractness) about accelerating learning and system thinking to tools and processes that help us to understand complexity, design better operating policies, and guide change in systems (concrete) from the smallest business to the planet as a world [16].

It is not easy to model the process of diffusion of a new technology. However, System Dynamics is an appropriate tool to model it. "System Dynamics is used for the analysis of policy and strategy, with a focus on business and public applications. Furthermore, System Dynamics is a perspective and set of conceptual tools that enable us to understand the structure and dynamics of complex systems" [16]. This modeling technique provides more concreteness to the abstract description of technology transitions.

System Dynamics models facilitate the development and testing of policies that lead to improved system behavior. It is very important to know that System Dynamics is not a tool designed to predict the future, but a tool to understand the present: to understand the variables that interact and how they can influence each other.

A causal loop diagram, as shown in Figure 1, is the tool used in System Dynamics to visualize the causal relations between variables in a system. A positive causal influence, symbolized by "+", means that if the influencing variable (which is at the arrow's origin) increases, then the influenced variable (which is at the arrow's head) increases. A negative causal influence, symbolized by "–", means that if the influencing variable increases then the influenced variable decreases.

With the causal loop diagrams we are able to visualize the feedback loops of a model. A feedback loop exists when a variable can influence its own value over time. In a positive, or amplifying, feedback loop, an increase in the variable will lead to a further increase of the variable over time. In a negative, or balancing, feedback loop an increase in a variable will lead to a decrease over time.

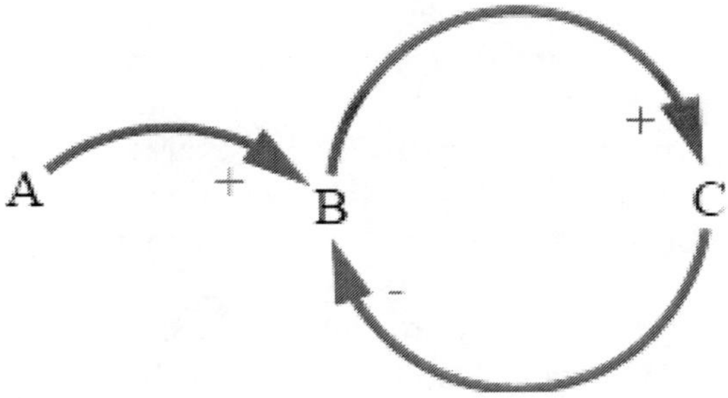

Figure 1. Casual loop diagram example.

3. GENERAL MODEL DESCRIPTION

Figure 2 shows a general vision of the model's blocks and the interrelations between them.

Initially, there is a demand for electric vehicles. The vehicles manufactured must satisfy this demand. For the study, the vehicles manufactured for the Spanish market are considered, not taking into account the country in which they are manufactured. As the number of manufactured vehicles increases, the electric vehicle park will also grow, which will clearly affect CO_2 emissions included in the environmental block and, subsequently, the policies options that the Government can use, either in the form of subsidies or special cargo rates, or an increase in the number of charging points, *etc.* The Government's policies will also be influenced by the environmental block, since the subsidies will be amended to reduce CO_2 emissions to the atmosphere.

The Government's policies will significantly influence the cost that the potential customer perceives when purchasing electric vehicles, due to subsidies, and the costs over the life of the vehicle by reduced overnight charge rates, *etc.* In addition, the Government's policies will also affect the attractiveness of electric vehicles for the consumer in comparison to a conventional vehicle. This is motivated by the advantages or disadvantages that buying an electric vehicle may have, such as the autonomy, the lack of charging points, the cost of recharging the electric vehicle, *etc*.

Both the cost of the electric vehicle and its attractiveness to the consumer are directly related to the evolution that the technology of electric vehicles will have in the coming years. This is because a very important part of the purchase price of electric vehicles is the battery, and one of the main factors of concern for potential buyers is the autonomy they have.

To close the loop, electric vehicle demand will evolve over time, influenced by the amount of money the buyer will have to pay (at the initial time of purchase and throughout the life of the vehicle), and its attractiveness.

It is interesting to note that, in global terms, the evolution of the technologies associated with electric vehicles will also be influenced by the demand for them. In this study, it is not considered in this way, since the demand for electric vehicles in Spain is very small in comparison to global demand, which will motivate technological advances in this field.

The same happens with other blocks, that is, the system being treated is not an isolated system. The global market for electric vehicles will influence the system, but for this study, we have considered various data as exogenous variables, *i.e.*, the value of these variables is determined by factors not included in the model being used. According to this, we can determine that the study covers worldwide developments.

4. BLOCK: ELECTRIC VEHICLE COST

Although all the blocks are treated in the full study, for this paper, only the block containing the electric vehicle costs and its comparison with the conventional vehicle are explained. This model's block analyzes the costs of both electric and conventional vehicles at time of purchase by the customer and during the vehicle's lifetime. To do this, data is considered as an evolution of oil prices and electricity, maintenance costs, *etc*.

To analyze the economic efficiency of electric vehicles, the total costs of ownership (TCO) of both vehicles (electric and conventional) are analyzed [17]. TCO describes costs over the lifetime of a vehicle: purchasing costs and costs of operation (annual costs). A comparison of the TCO of electric vehicles with the reference vehicles provides indications of the economic efficiency of the former. The battery plays a major role in the TCO, so battery costs are looked at in detail.

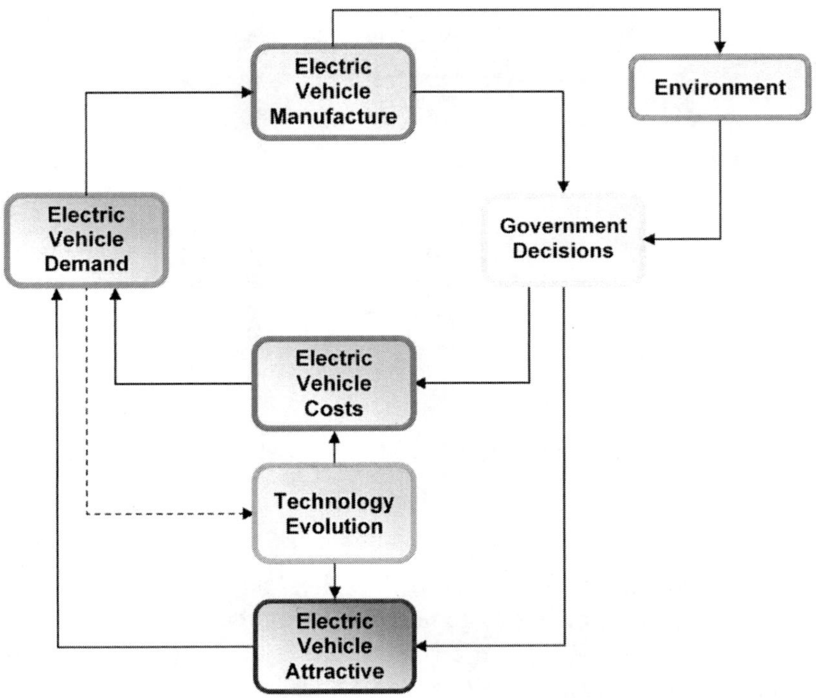

Figure 2. General model schema.

The model has been divided into four sections: the purchase cost of the electric vehicle, the cost of the electric vehicle during its life, the purchase cost of the conventional vehicle, and the cost of operation of the conventional vehicle during its life. The total cost of electric vehicles consists of the sum of the purchase cost and the running costs of electric vehicles. In the same way, the total cost of conventional vehicles consists of the sum of the purchasing cost and the running costs of conventional vehicles. Figure 3 shows a summary of the overall functioning of the block.

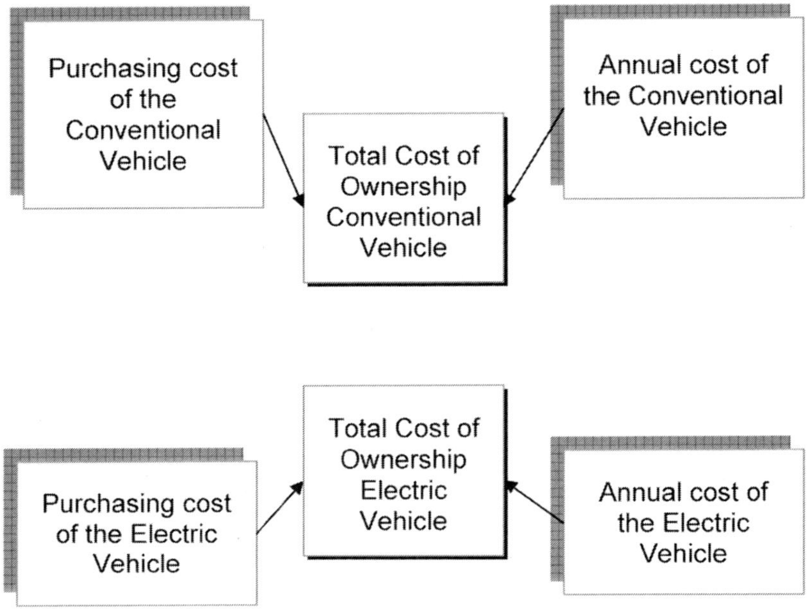

Figure 3. Schema of the electric vehicle block.

The simplest section corresponds to the purchasing cost of the conventional vehicle by the customer. The conventional vehicles that can potentially compete with electric vehicles are relatively compact, so that, based on statistical data and purchasing trends in Spain, a reference price has been set for the conventional vehicle.

Once the reference value of the conventional vehicle has been set, the purchase cost of electric vehicles is determined by the reference value of the conventional vehicle, adding some kind of costs, but also eliminating others. Figure 4 shows a simplified diagram of the electric vehicle.

The most significant additional costs of electric vehicles will be the necessary wiring and charger for charging the vehicle both slowly and rapidly [2], the different motor drive required and, of course, the cost of the batteries, which is the most significant contributor.

According to the available information [2], the extra charger and wiring costs in Spain will cost about 480 €, while the cost of the motor drive will be 160 €. In contrast, the motor simplification leads to a reduction of product cost. It is estimated that a gasoline engine will cost about 2,400 €, while the electric motor will only cost 1,200 €, so there is an interesting reduction.

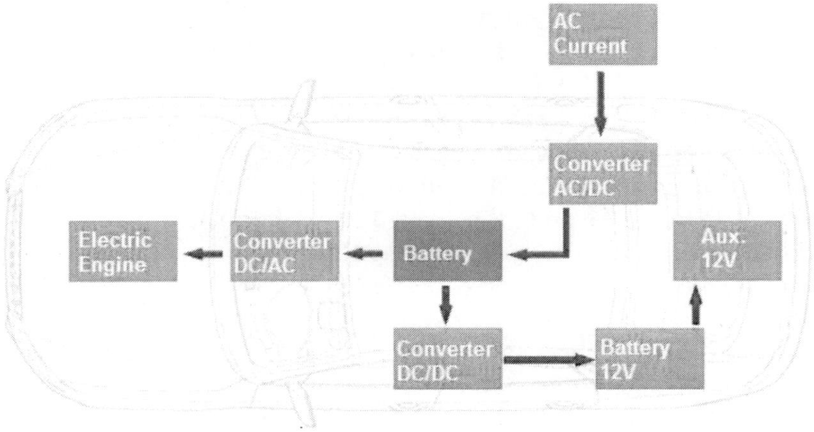

Figure 4. Electric vehicle schema.

Moreover, the electric motor requires an additional component: an engine control system that allows the required torque or speed to be adjusted for every load point. The cost of this control is now about 1,020 € but a significant reduction is expected to result from mass production, with the price reaching 70% of its current value [18].

In electric vehicles, an additional component called a DC/DC converter is also required. This is used to convert the voltage from 100 V to 12 V, which is necessary for the auxiliary systems required for the vehicle. The cost will depend heavily on the input voltage, but should be around 300 € [18].

In addition to this converter, an AC/DC converter, used to recharge the battery by connecting it to an external AC power, is also needed. The cost of this converter is approximately 400 € [18].

One of the key factors in the evolution of the purchasing price of electric vehicles will be the reduction in the battery costs in the coming years. To predict these developments, there is a model which includes forecasted values of the reducing cost of the kWh, and the evolution that the battery capacity will undergo in the coming years in both optimistic and pessimistic scenarios [19], developed in the Technology Block, mainly using works by Kromer and Heywood [20], Duvall [9] and Kurani [13].

To calculate this, it is also necessary to know the battery's capacity and the manufacturing cost of the battery. For the cost estimate, we use a graph of the expected price per kWh reduction over the next 10 years. To do this, we use the following reference data [19] from the Technology Block.

As can be seen in the graph (Figure 5), there are several scenarios for the cost of batteries, but what is quite clear is that they provide a very significant decrease. So far, the rate of decline in the cost of the batteries has been between 6% and 8% per year [21]. Many analysts predict that the price will continue to fall over the next 10 years in the same way, due to increased production volumes.

A cost that has also been considered for this study is the upgrading cost of the garage, in appropriate cases, to be able to recharge the battery there. Taking into account the potential population living in apartment buildings, in single-family houses, and the Spanish law, the upgrading cost will vary significantly.

According to a RACE (Royal Automobile Club of Spain) study [22], 80% of the population would choose the nighttime to charge electric vehicles, so to simplify the model it is assumed that in these early years of commercialization of electric vehicles a customer acquiring an electric vehicle will also charge it in his garage during the night.

At this point it is very important to know whether the available garage is individual or community property, because, though the charging systems are similar, the method for calculating electricity consumption differs considerably in each case. On the one hand, in an individual garage, all electricity consumed (either by charging an electric vehicle or by having a computer connected, or air conditioner for example) will be billed to a single consumer. On the other hand, in the community garage, it is necessary to discriminate the source of consumption in order to know whom it will be charged to.

According to data from the INE (National Institute of Statistics), 28% of the population lives in a single-family house, while 72% live in a flat. While not 100% of the population has a garage, this study has ignored that fact, since, according to all surveys [22], the potential client owns a garage.

In a single-family house garage, it is simply necessary to have a 230 V and 16 A household outlet, for which the cost has been estimated at approximately 100 € for a basic installation in an environment with electric service available.

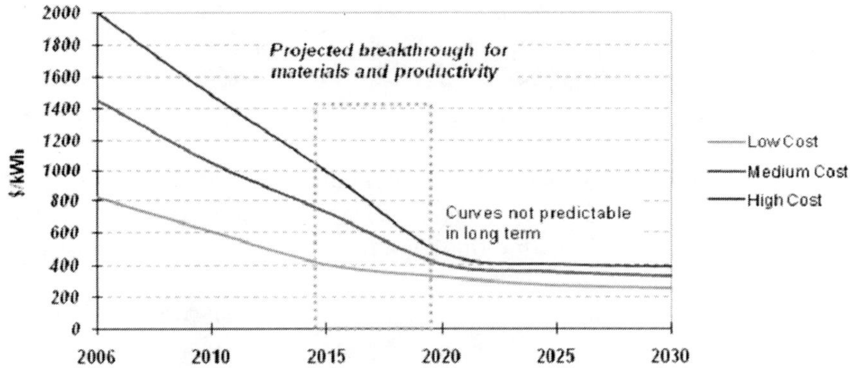

Figure 5. Forecast of the battery price [19].

In the case of collective garages, the required system is slightly more complex. While the law already authorizes and provides facilities in such collective garages (Law 49/1960, of July 21st), in this case, as well as having a socket of similar dimensions to those of an individual garage, a counter to discriminate the source of consumption is also needed. Thus, the residents pay the total electricity consumed, then that cost is distributed among users of existing electric vehicles in proportion to their consumption. However, in anticipation of widespread recharging of electric vehicles in collective garages, several brands of electric circuits are beginning to market solutions for collective garages such as charging system monitoring interfaces, or in its simplest case, individual recharging points activated by an identification system to eliminate potential vandalism.

In view of the electrical systems that must be installed in the collective garage, it is logical to expect that the cost of installation will be very superior to that in the individual garage. Taking into account actual budgets, and also taking into account the need for the implementation of a certification for it to be a new installation, the estimated average cost for charging an electric vehicle in a community garage is 900 €.

A final factor to consider in the purchase price of electric vehicles is the subsidies that the government can provide at the specific time of purchase. For this calculation, we use three interrelated variables: the value of the subsidy, the budget that the IDAE (Spanish Energy Institute) can spend on subsidies for the purchase [1] and, finally, the money already being spent on subsidies.

Currently, 7,000 Euros has been taken as the purchase subsidy [1], which is the average subsidy for this type of vehicle in Spain. The government budget is 240 million Euros in 2011 and 2012. For 2012, no data or estimations of subsidies have yet been delivered.

Once the purchasing costs for both vehicles have been established, it is now necessary to determine the running costs, which in both cases are mainly the cost of fuel (and its consumption) and the cost of vehicle maintenance.

For the conventional vehicle, the analysis of the annual fuel consumption begins with the average consumption of an internal combustion vehicle today in Spain, taking into account the average consumption of diesel and gasoline vehicles, as well as the percentage of population using each kind of vehicle. It is considered that the average consumption of a gasoline engine is 8 liters per 100 km (20.4 miles per gallon), while the average consumption of a diesel vehicle is 6 liters per 100 km (39.2 miles per gallon) [23]. Given that in Spain 53% of vehicles are gasoline and 47% diesel [24], the current average consumption of an internal combustion vehicle is the product of the consumption of gasoline vehicles times their percentage product, added to the consumption of diesel vehicles multiplied by their percentage. After obtaining some reference data, some improvements in the efficiency of the engines that will be developed in coming years have been added as well as aerodynamic improvements that will dynamically change vehicle consumption [2,18].

Estimating how oil prices will evolve over the next 10 years is a task of great complexity with a high rate of failure. The price of oil depends on many factors. There are few reserves in the World and it is a scarce commodity (referring to what can be extracted with current technology, and whose extraction would be profitable at current oil prices).

Another important factor to consider is the large scale financial speculation, apart from the ongoing geo-political conditions (oil prices may rise due to political situations involving producing countries).

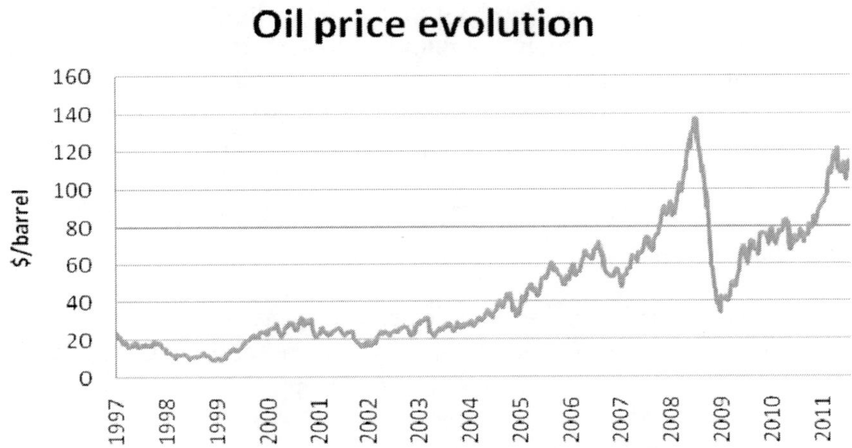

Figure 6. Oil price evolution [25].

Figure 6 shows the evolution of oil prices in recent decades. It shows the great variability of the price and the mathematical complexity of this development model [25]. However, to obtain a more realistic model, the fixed oil price (the value at the beginning of 2010) has been included [26]. The forecast values in oil prices in dollars/barrel are:

Table 1. Oil prices forecasted [26].

YEAR	2010	2011	2012	2013	2014	2015	2016	2017	2018	2019	2020
PRICE	70.30	73.06	79.41	85.74	90.91	94.52	98.23	101.23	104.41	106.47	108.28

The tax regulations fixed by Spanish legislation need to be added to this oil price, as well as transport and the profit margin of gas stations. According to data from the "Market Observatory for Energy" [26], indirect taxes for gasoline are 436.47 € per 1000 liters, while for diesel fuels they are 340.36 € per 1000 liters. The V.A.T. tax in Spain is at 18%, and the profit margin of gas stations in Spain is about 25%.

Finally, we estimated the maintenance costs that conventional vehicles will have on an annual basis. For this, we estimated the maintenance costs that a conventional vehicle has over its useful lifetime, and this estimation has been done taking into account the number of kilometers a vehicle does on average before it needs to be replaced. The value will be set by external studies that determine the value as 0.03 € per kilometer covered [11,27]. If this value is multiplied by the average annual kilometers traveled by a vehicle, this will have, as a result, an economic value which is the amount that the customer spends on an annual basis on conventional vehicle maintenance.

For the running costs of the electric vehicle, the schema is similar to that of the conventional vehicle: There is an annual cost due to vehicle consumption and other costs due to maintenance.

The annual cost derived from the electric vehicle consumption is calculated based on the price of electricity, but it is necessary to take into account various considerations: the number of users that will have a suitably equipped garage, the number of users that will recharge by night, by day, the driving patterns [21], *etc.*

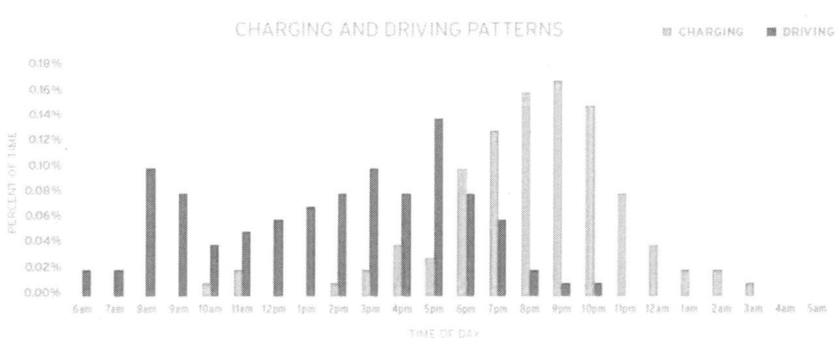

Figure 7. Driving patterns [21].

The first consideration is to assume that most users with electric vehicles will have a garage (as seen before) and most of them (80% [22]) will recharge the car at night. This value is also supported by studies in pilot positions analyzing the patterns of use of the electric vehicle and its recharge rate. Figure 7 shows the results of a study of "The Idaho National Laboratory" [21]. It can be clearly seen how the majority of recharges are done during a period when the vehicle is not in use, at nighttime.

According to this, the prices of recharge during the day, the evening and at the reduced rate called "Super Off-peak Tariff", given by the Government of Spain, were studied. The last rate is designed to support the charges of electric vehicles made during the night, and thereby improve the recharge curve as explained above, as reflected in the "Strategy to Promote Electric Vehicles in Spain" [28]. Since there is still no information on this rate, we estimated a maximum cost per kWh that would give this rate-calculated value at 50% of the cost of electricity during the night. This value has been taken in a somewhat arbitrary way, since, as mentioned above, there is no information yet on this rate, but it is considered that this value of kWh could achieve the results anticipated at this rate.

With the values of all rates and the estimations of the potential users of each rate, the average price of electricity has been obtained. As happened with oil prices, obtaining an estimate of the price of electricity in Spain is extremely difficult, but it has been necessary to make this estimate for a more realistic model. Data based on from the values from recent years cannot be extrapolated, given the continuous modification of the rates due to a tariff deficit of the electrical system.

The tariff deficit is the difference between the actual cost of electricity production (both production and transport and distribution) and the fixed rate. This generates a debt that is accumulated every year. Because of this, the fluctuation in prices is difficult to quantify. According to this, the approach indicated by the European Commission for Energy has been used, given by the curve shown in the Figure 8.

Finally, it is necessary to estimate the maintenance costs in the same way as for conventional vehicles, that is, relative to the number of kilometers a vehicle does on average before it needs to be replaced. In this case, the value will be set by external studies that determine the value as 0.02 € per kilometer [11,30].

By multiplying this value by the average annual miles traveled by a vehicle, you get the economic value that the customer spends annually to maintain the electric vehicle. As can be seen, the maintenance cost per kilometer is significantly lower compared to the conventional vehicle (33% lower). This is because electric motors do not have a large variety of working parts, which means that a continuous maintenance, as exhaustive as in internal combustion engines, is not necessary. In fact, the electric motor has one moving part, the shaft, which is very reliable and requires little or no maintenance. Another important factor is the elimination of oil, which implies that it is not necessary to make periodic oil changes. In the electric vehicle, drive systems do not exist, as nor do fuel pumps or injection, which also minimizes maintenance [8]. The fact that the electric motor is simpler than the internal combustion engine reduces the risk of breakdowns.

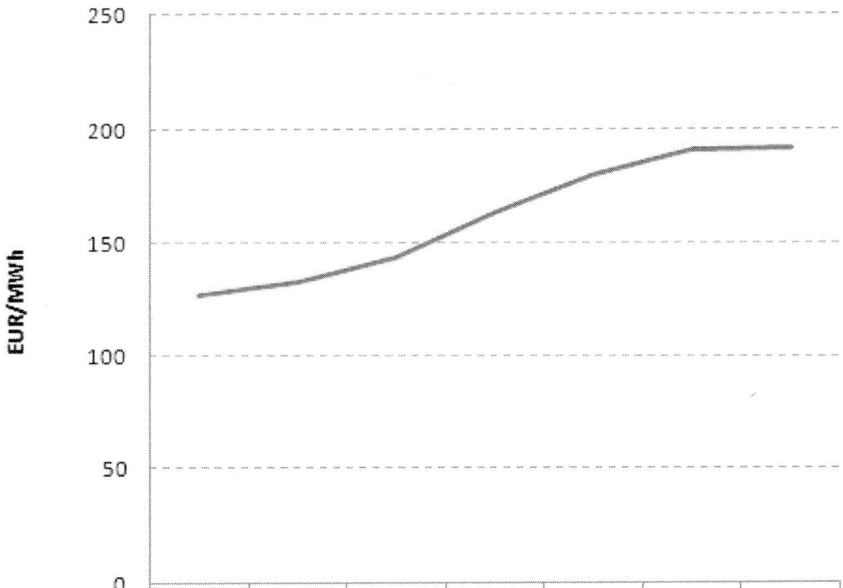

2000	2005	2010	2015	2020	2025	2030
127	133	144	164	180	191	192

Figure 8. Electricity price forecast [29].

To summarize, we can determine that the TCO is described by the following expression:

$$\text{TCO} = \text{Purchasing cost } (y) + \sum_{t=y}^{2020} (\text{Fuel cost } (t) + \text{Maintenance cost}) \tag{1}$$

where "*y*" is the year of purchase. As explained above, the annual costs for both kinds of vehicles are determined by fuel costs and maintenance costs from the vehicle purchasing year until the year 2020, the year when the simulation ends. Obviously, these costs are variable in time due to variations in the fuel prices previously explained, as well as the number of kilometers traveled on average each year.

Once all input data have been determined and modeled, the output data is the number of years that the potential clients take to recover their investment, comparing how long it takes for it to be more profitable to buy an electric vehicle than a conventional vehicle. This value will depend on the timeframe when purchasing the vehicle.

According to the number of years for repayment, the purchase of electric vehicles, compared to conventional vehicles, will have a different appeal to the potential customer, resulting in a percentage of the population that will potentially want to buy an electric vehicle in Spain.

This result, added to the rest of the results from the other blocks of the model, enables us to obtain the estimated number of vehicles that could run on Spanish roads in the coming years under our assumptions.

It is important to note that, although the economic is the main factor when purchasing a new vehicle, there is no direct correlation between the price of the vehicle and the number of existing electric vehicles on the roads. For the full study and the final results, other factors have been used, such as the autonomy of electric vehicle, advertisements and information, recharging points and “word of mouth”.

6. RESULTS

As explained before, the scenarios to analyze are related with the economic factor, keeping the baseline scenario for the other variables not analyzed in this paper. The first scenario to be analyzed corresponds to the baseline model, based on data that we know today for Spain; that is, a forecast model in terms of the likely evolution of the batteries and the non-implementation of battery exchange stations [31] and with the estimated average battery cost. We could say that this is the reference state. To combine these data, it is assumed that government subsidies are known until this time, *i.e.*, 7,000 € for the purchase. As for the cost of the battery, according to the average forecast in 2020, the cost per kWh will be close to 243 € [21], implying that a full battery will cost about 6,500 €. According to the model and the average forecasts in all fields, we find that the number of vehicles in 2020 would be around 700,000 (Figure 9).

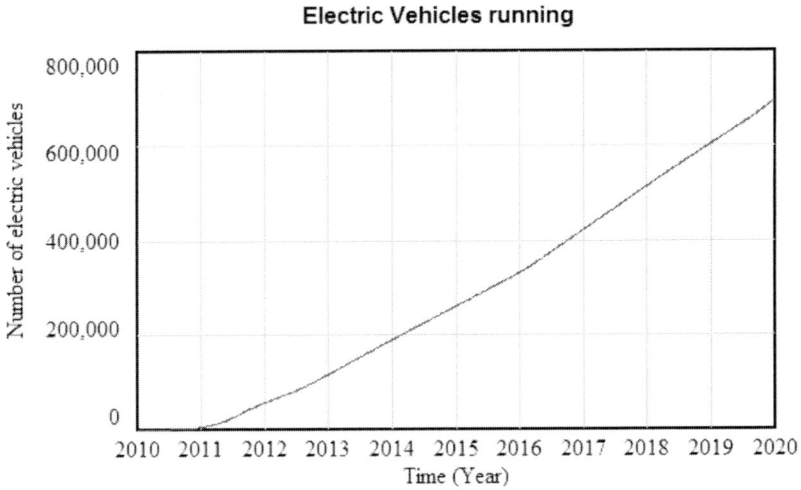

Figure 9. Electric vehicles on Spanish roads (the authors' calculations).

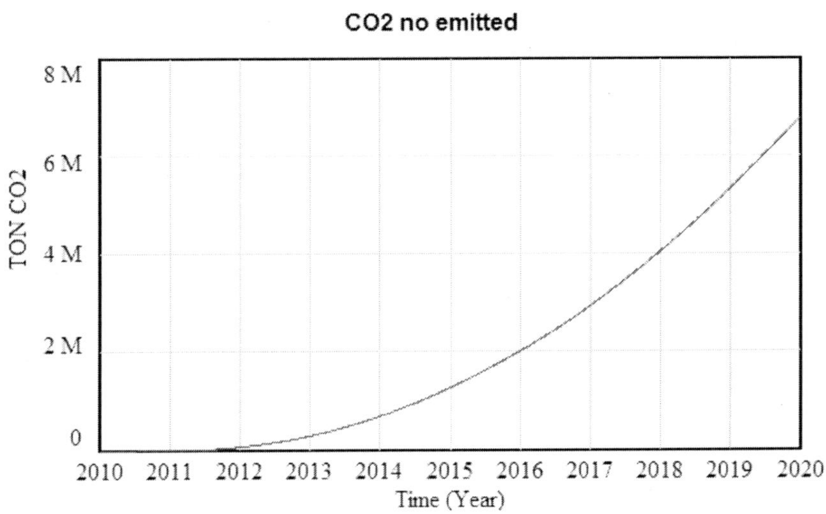

Figure 10. TON of CO_2 not emitted (the authors' calculations).

This figure corresponds to a number of CO_2 tons not emitted to the atmosphere of more than 6 million (Figure 10). To calculate the CO_2 not emitted, the replacement of a conventional vehicle by an electric vehicle is assumed. In this first stage of the electric vehicle commercialization, it is only expected to reach the "early market", that is, the market formed by the "innovators" such as car enthusiasts and environmentalists [30]. Due to the small size of this group (less than 2.5% of global demand), the so-called "rebound effect" due to the temporal space covered by the study is not expected [32].

Although the electric vehicle obtains a reduction in CO_2 emissions because it does not generate emissions during its use due to the electric engine, there are emissions generated in the production of the electrical energy used by the vehicle. Accordingly, there is an emission displacement, but as mentioned above, the total energy balance (Well-To-Wheel) still indicates a significant reduction in CO_2 emissions [33].

The following scenarios are designed to study the evolution of the cost of the battery. As seen before, the cost of the battery is decisive in the evolution of electric vehicle demand. For the analysis of this battery development, two scenarios are considered in addition to the baseline: the optimistic scenario with an annual decrease of 10% in kWh of the battery (corresponding to a major evolution of electric vehicle sales worldwide, making the cost of the battery dramatically lower due to mass production), and one pessimistic, with an annual decrease of 4% in the price of the battery (this scenario represents a poor use of electric vehicles, where mass production of batteries has failed to take off sufficiently in order to achieve a significant reduction in the price). The baseline scenario is a decrease of 7%. Figure 11 shows the cost of the battery under the various scenarios.

In these three cases, the electric vehicle fleet evolves as follows (Figure 12):

It can be seen that, in 2013 and the following years, there is a clear difference between the various scenarios. This is because the battery capacity in these early years of commercialization of electric vehicles is relatively small (22 kWh). The value of kWh between scenarios in 2013 ranged between values of 364 and 442 €, *i.e.*, about 80 € of difference, which, increased by 22 kWh, is approximately 1,700 €. As the battery increases in capacity, this price difference becomes more significant. This can clearly be seen in the graph of the average price of electric vehicles with the current values of the facilities (Figure 13).

As seen before, the price of the electric vehicle grows so much during 2012 due to the end of subsidies currently approved by the Spanish government. If government support in the form of a subsidy for purchasing electric vehicles is stopped, the targets will not be achieved. Figure 14 shows the optimistic scenario of reducing the cost of the battery in relation to government objectives.

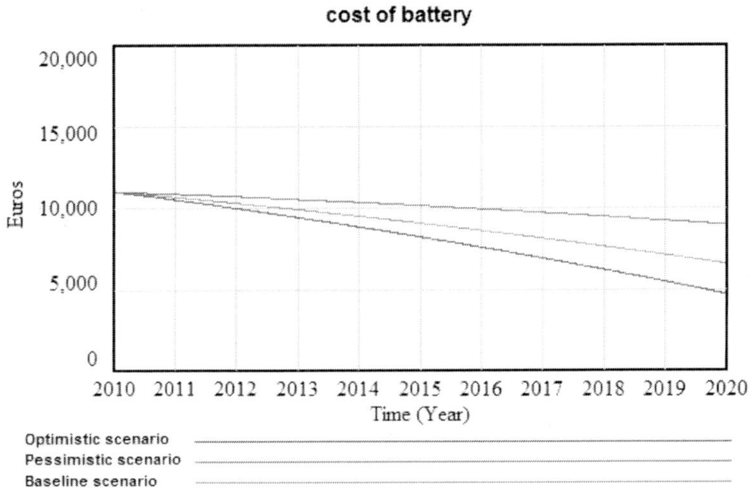

Figure 11. Forecast of battery cost in the optimistic, pessimistic and baseline scenarios (own calculations).

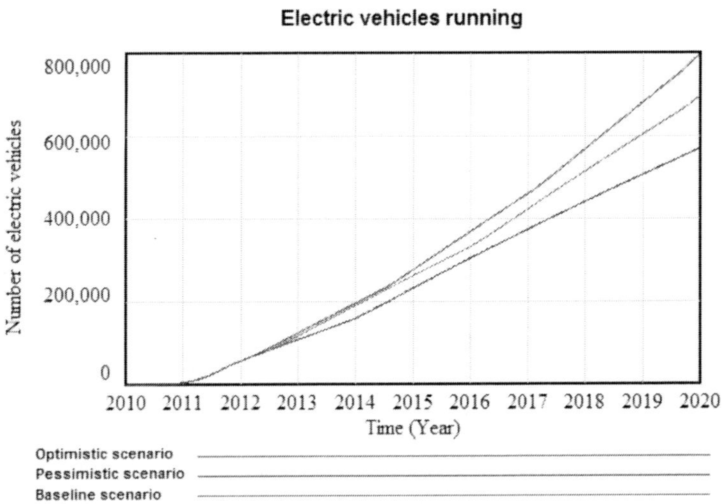

Figure 12. Electric vehicles on roads for the optimistic, pessimistic and baseline scenarios (the authors' calculations).

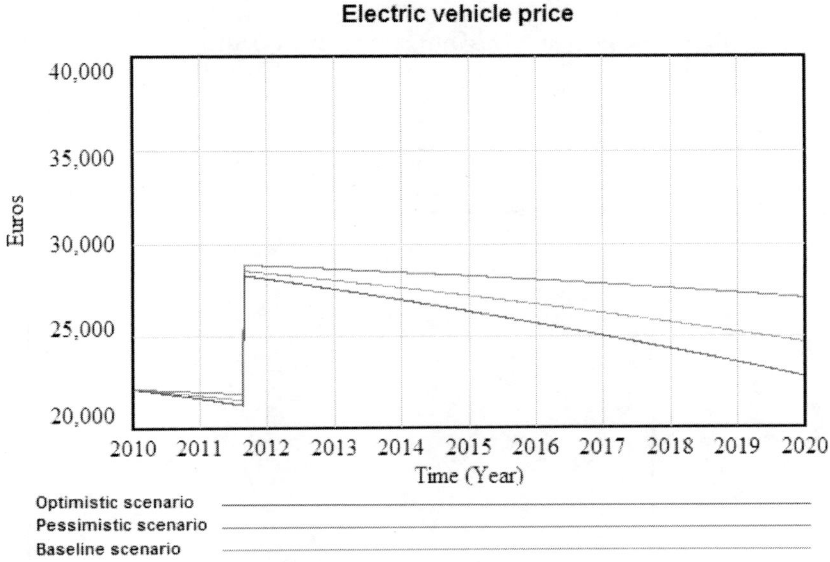

Figure 13. Electric vehicle price for the optimistic, pessimistic and baseline scenarios (own calculations).

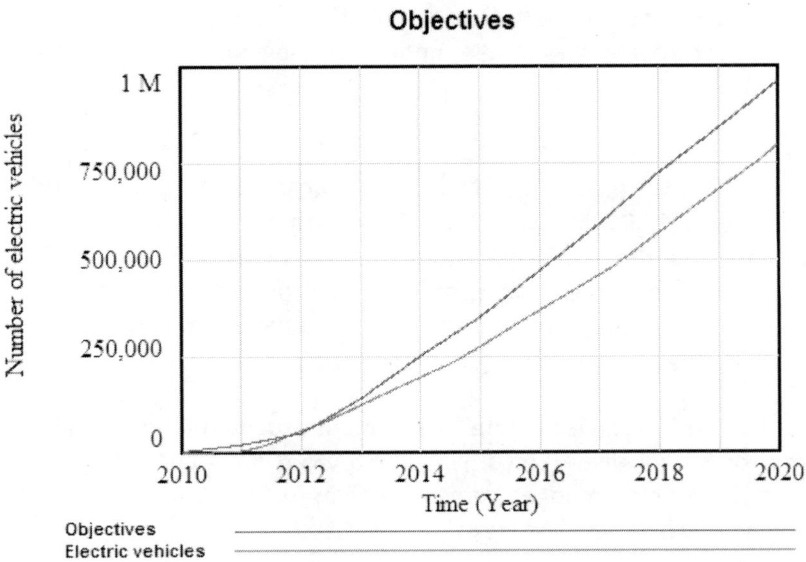

Figure 14. Electric vehicle fleet compared to the Government target for the optimistic scenario without subsidies in 2013 (the authors' calculations).

If we keep the subsidies of 7,000 € constant in time for the optimistic scenario, the following results are obtained (Figure 15).

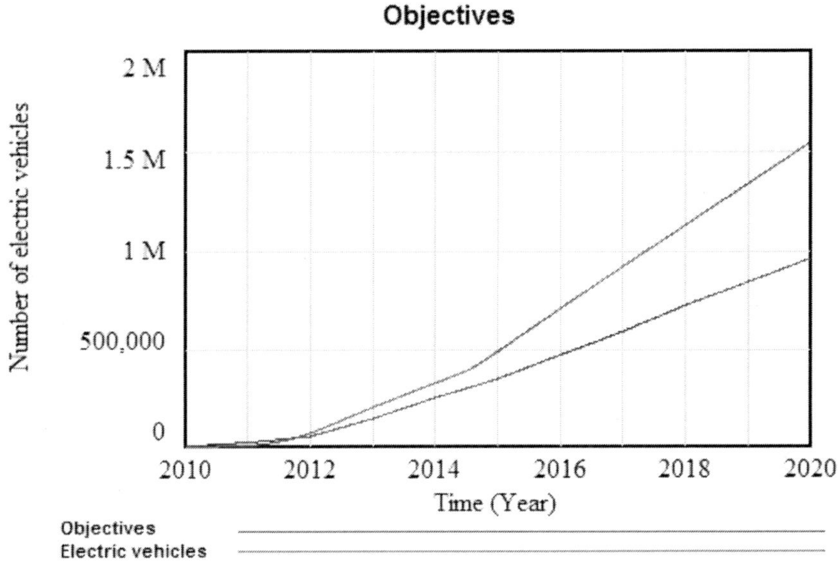

Figure 15. Electric vehicle fleet compared to the Government target for the optimistic scenario with subsidies until 2020 (the authors' calculations).

In this case, the results are well above the targets set, but at an exorbitant cost (close to 14,000 million Euros). By performing various simulations, we can determine that in the optimistic scenario of reducing the cost of the battery, with a subsidy value of 2,000 € from 2012, the government achieves the objectives until at least 2016, and there is a slight difference in subsequent years, only converging after 2020.

For 2,000 € of subsidy, the total government expenditure will be slightly in excess of 3,000 million Euros, which is a value quite consistent with the current government budget (remember that, for the years 2011 and 2012, the budget was 240 million Euros) (Figure 17).

The case of the pessimistic scenario is quite different from the optimistic one, given the value of the subsidies necessary to achieve the objectives. When using the estimated value of 2,000 € for the optimistic scenario, we find that the goal is met until 2014, but as the battery capacity increases over time, demand for electric vehicles will decrease, since they will lose competitiveness against conventional vehicles, due to high additional costs. In this case, the optimal value of the subsidies, after making several simulations, were obtained for values between 4,500 € and 5,000 €, implying a cost to the government of nearly 6,000 million Euros (Figure 18).

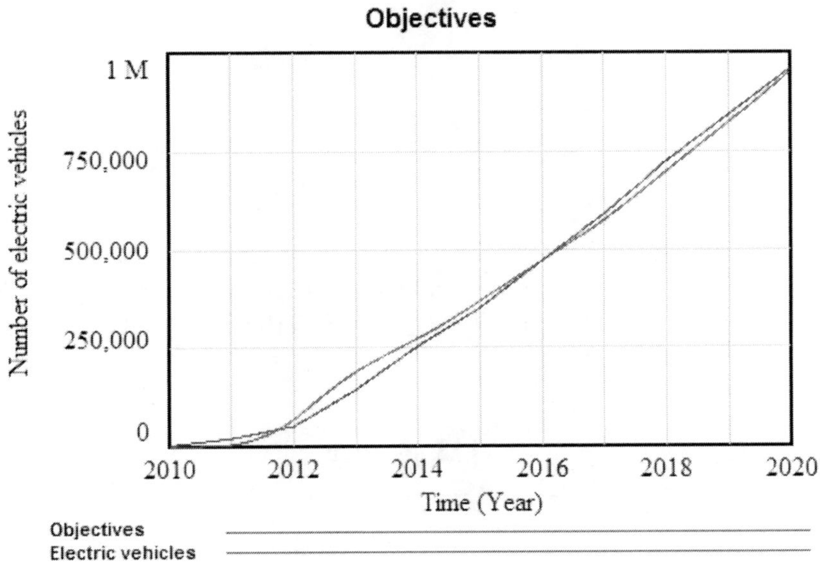

Figure 16. Electric vehicle fleet compared to the Government target for the optimistic scenario with optimal subsidies until 2020 (the authors' calculations).

In view of the results, which can be determined taking into account the decrease in the cost of the kWh battery that will exist in coming years, the value of the subsidies to be given for the purchase of electric vehicles will range from 2,000 € in the optimistic case and up to 5,000 € in the pessimistic case.

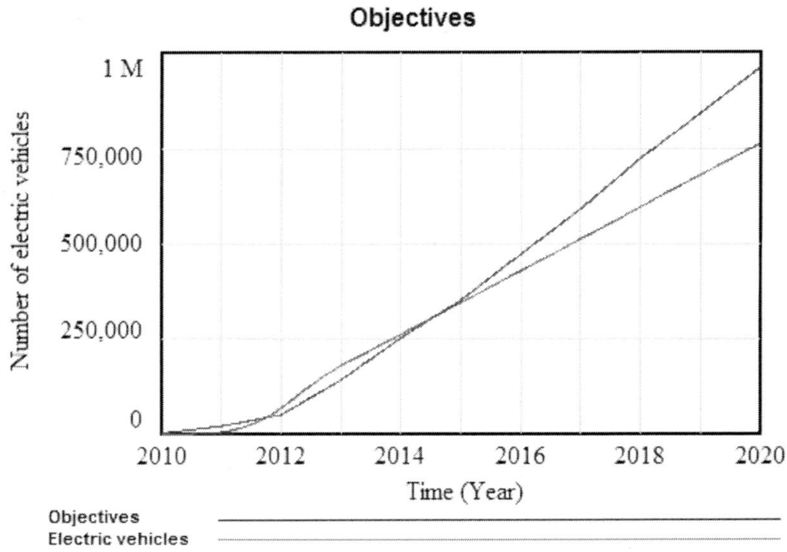

Figure 17. Electric vehicle fleet compared to the Government target for the pessimistic scenario without subsidies until 2020 (the authors' calculations).

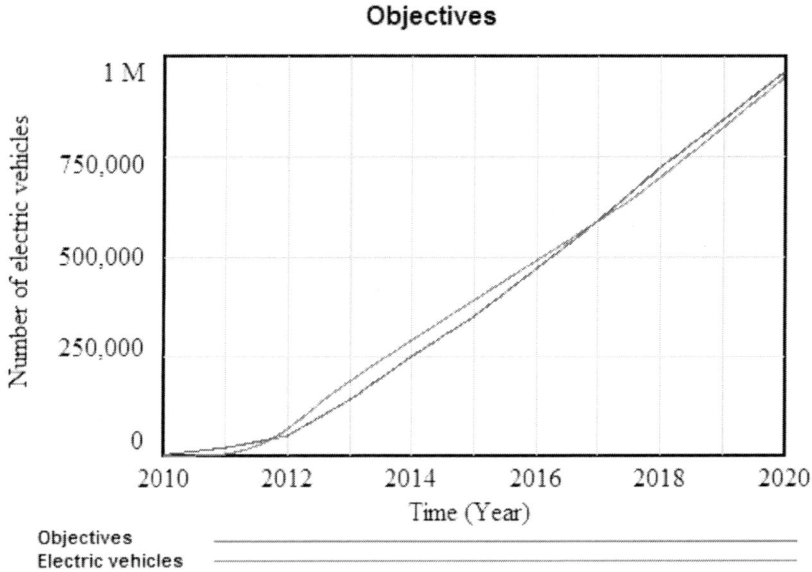

Figure 18. Electric vehicle fleet compared to the Government target for the pessimistic scenario with the optimal subsidies until 2020 (the authors' calculations).

7. CONCLUSIONS

The proposed model based on System Dynamics, provides a structure and framework to understand the behavior of the electric vehicle market according to the behavior of the main variables involved and the different scenarios. The key factors to integrate energy policies with market and customer criteria have been shown and quantified. The simulations help to understand the problem and forecast the evolution of electric vehicles in Spain under different hypotheses.

According to the simulation results, regardless of technological developments that may occur in the coming years, if government subsidies are limited to 2012, the current targets set for 2020 concerning the number of electric vehicles circulating on Spanish roads will not be achieved.

In view of the analysis of the results of the different scenarios, for a good implementation of electric vehicles, a major government outlay to subsidize the purchase will be needed. The acquisition of electric vehicles compared to conventional ones has a cost which is not always accepted by the consumer (although it will be compensated for in the use of the vehicle). In none of the scenarios discussed (even in the most optimistic scenario) are the desired results obtained if subsidies to them are the potential buyer are discontinued. While it is true that the value of the subsidies provided today will not be necessary in the near future, being able to significantly reduce this amount appears to be necessary, at least until 2020.

In fact, the economic factor is so important that, for the potential client, the scenarios of the evolution of vehicle autonomy will not prevail over the economic aspect. Moreover, according to the results of the simulations, the customer will prefer a vehicle with less autonomy, but at a lower cost. Therefore, although there will be significant technological improvements, an extra financial contribution will still be needed when the vehicle is purchased.

Briefly, if the Spanish government maintains its goals for electric vehicles and there is no unexpected technological leap, new economic or financial policies will be needed to support the electric vehicle market. The quantitative results about possible subsidies are only guideline values and they are conditioned by the proposed hypothesis and scenarios.

REFERENCES

1. Ministerio de Industria, Comercio y Turismo; IDAE. Proyecto Movele. Proyecto piloto de demostración de viabilidad del vehículo eléctrico. 2009. Available online: http://www.idae.es/ index.php/mod.documentos/mem.descarga?file=/documentos_Cuadriptico_MOVELE_Final_fac4a8ee.pdf (accessed on 6 July 2010).

2. Bandivadekar, A.; Bodek, K.; Cheah, L.; Evans, C.; Groode, T.; Heywood, J.; Kasseris, E.; Kromer, M.; Weiss, M. On the Road in 2035: Reducing Transportation's Petroleum Consumption and GHG Emissions. Massachusetts Institute of Technology: Cambridge, MA, USA, 2008. Available online: http://web.mit.edu/sloan-auto-lab/research/beforeh2/otr2035/ On%20the%20 Road% 20in%202035_MIT_July%202008.pdf (accessed on 7 May 2010).

3. European Council for Automotive R&D. Well-to-Wheels Analysis of Future Automotive Fuels and Powertrains in the European Context. 2006. Available online: http:// co2star.eu/publications/ Well_to_Wheels_ Report_EU.pdf (accessed on 28 July 2010).

4. Granovskii, M.; Dincer, I.; Rosen, M.A. Economic and environmental comparison of conventional, hybrid, electric and hydrogen fuel cell vehicles. *J. Power Sources* **2006**, *159*, 1186–1193.

5. Walther, G.; Wansart, J.; Kieckhäfer, K.; Schnieder, E.; Spengler, T.S. Impact assessment in the automotive industry: Mandatory market introduction of alternative powertrain technologies. *Syst. Dyn. Rev.* **2010**, *26*, 239–261.

6. Train, K.E.; Winston, C. Vehicle choice behavior and the declining market share of the US automakers. *Int. Econ. Rev.***2007**, 1469–1496.

7. Borenstein, S. Cost, Conflict and Climate: US Challenges in the World Oil Market. 2008. Available online: http://ucei. berkeley.edu/PDF/csemwp177.pdf (accessed on 2 April 2011).

8. Werber, M.; Fischer, M.; Schwartz, P. Batteries: Lower cost than gasoline? *Energy Policy* **2009**, *37*, 2465–2468.

9. Duvall, M.; Browning, L.; Kalhammer, F.; Warf, W.; Taylor, D.; Wehrey, M.; Pinsky, N. *Advanced Batteries for Electric-Drive Vehicles: A*

Technology and Cost-Effectiveness Assessment for Battery Electric Vehicles, Power Assist Hybrid Electric Vehicles, and Plug-in Hybrid Electric Vehicles; Electric Power Research Institute: Palo Alto, CA, USA, 2004. Available online: http://www. spinnovation.comsn/ Batteries/ Advanced_Batteries_for_Electric-Drive_Vehicles.pdf (accessed on 9 August 2011).

10. Graham, R. Comparing the Benefits and Impacts of Hybrid Electric Vehicle Options. Electric Power Research Institute: Palo Alto, CA, USA, 2001. Available online: http://www.ourenergypolicy.org/docs/9/ Comparing_Hybrid_Electric_Vehicle_Options.pdf (accessed on 10 August 2011).

11. Neenan, B. *Characterizing Consumers' Interest in and Infrastructure Expectations for Electric Vehicles: Research Design and Survey Results*; Electric Power Research Institute: Palo Alto, CA and Southern California Edison, Rosemead, CA, USA, 2010. Available online: http://my. epri.com/ portal/server.pt? Abstract_id= 000000000001021285 (accessed on 10 August 2011).

12. Kurani, K.; Axsen, J.; Caperello, N.; Davies, J.; Stillwater, T. *Plug-In Hybrid Electric Vehicle (PHEV) Demonstration and Consumer Education Outread, and Market Research Program*; Institute of Transportation Studies, University of California: Davis, CA, USA, 2009. Available online: http://pubs. its.ucdavis.edu/do wnload_pdf.php?id=1310 (accessed on 6 August 2011).

13. Kurani, K.; Axsen, J.; Burke, A. *Batteries for Plug-in Hybrid Electric Vehicles (PHEVs): Goals and the State of Technology circa 2008*; Institute of Transportation Studies, University of California: Davis, CA, USA, 2008. Available online: http://pubs.its.ucdavis.edu/download_ pdf.php?id=1169 (accessed on 6 August 2011).

14. Lumbreras, J.; Valdés, M.; Borge, R.; Rodríguez, M.E. Assessment of vehicle emissions projections in Madrid (Spain) from 2004 to 2012 considering several control strategies. *Transp. Res. Part A* **2008**, *42*, 646–658.

15. Geels, F.W. From sectoral systems of innovation to socio-technical systems Insights about dynamics and change from sociology and institutional theory. *Res. Policy* **2004**, *33*, 897–920.

16. Sterman, J.D. *Business Dynamics: Systems Thinking and Modeling for a Complex World*; Irwin McGraw-Hill: Burr Ridge, IL, USA, 2000.

17. Brickert, S.; Kuckshinrichs, W. Electromobility as a Technical Concept in an Ecological Mobility Sector? An Analysis of Costs. Proceedings of the 9th International Conference of the European Society for Ecological Economics (ESEE 2011), Istanbul, Turkey, 14–17 June 2011; Available online: http://www.esee2011.org/registration/fullpapers/esee2011_54ff9e_2_1307627504_8295_2080.pdf (accessed on 6 August 2011).

18. Blesl, M.; Bruchof, D.; Hartmann, N.; Özdemir, D.; Fahl, U.; Eltrop, L.; Voß, A. *Entwicklungsstand und Perspektiven der Elektro-mobilität*; Universität Stuttgart Institut für Energiewirtschaft und Rationelle Energieanwendung: Stuttgart, Germany, 2009. Available online: http://elib.unituttgart.de/opus/volltexte/2010/5218/pdf/Elektromobilitaet_Endbericht_20100322.pdf (accessed on 7 August 2011).

19. Hensley, R.; Knupfer, S.; Pinner, D. McKinsey Quarterly, nº3. Electrifying Cars: How Three Industries Will Evolve. Available online: http://www.mckinseyquarterly.com/Electrifying_cars_How_three_industries_will_evolve_2370(accessed on 7 April 2010).

20. Kromer, M.A.; Heywood, J.B. *Electric Powertrains: Opportunities and Challenges in the U.S. Light-Duty Vehicle Fleet*; Laboratory for Energy and the Environment, Massachusetts Institute of Technology: Cambridge, MA, USA, 2007. Available online: http://web.mit.edu/sloan-auto-lab/research/beforeh2/files/kromer_ electric_powertrains.pdf (accessed on 7 April 2010).

21. Electrification Coalition. Electrification Roadmap. Revolutionizing Transportation and Achieving Energy Security. Available online: http://www.electrificationcoalition.org/sites/default/files/SAF_1213_EC-Roadmap_v12_Online.pdf(accessed on 7 April 2010).

22. Royal Automobile Club of Spain Website. Available online: http://www.race.es/ (accessed on 15 April 2010).

23. Club Español de la Energía; Instituto Español de la Energía. Cuadernos de Energía, nº26, octubre de 2009. Available online: http://www.enerclub.es/files/frontAction.do?action=getFile&fileID=1000056208 (accessed on 4 June 2010).

24. Traffic Department website. Available online: http://www.dgt.es/portal/ (accessed on 25 April 2010).

25. U. S. Energy Information Administration website. Available online: http://www.eia.gov (accessed on 2 February 2011).

26. U.S. Energy Information Administration. Annual Energy Outlook 2010. Available online: http://www.eia.gov/oiaf/archive/aeo10/pdf/0383%282010%29.pdf (accessed on 28 July 2011).

27. Pistoia, G. *Electric and Hybrid Vehicles: Power Sources, Models, Sustainability, Infrastructure and the Market*; Elsevier: Amsterdam, The Netherlands, 2010.

28. Ministerio de Industria, Comercio y Turismo. *Estrategia integral para el impulso del vehículo eléctrico en España*; Ministerio de Industria, Comercio y Turismo: Madrid, Spain, 2010.

29. European Commission Directorate-General for Energy and Transport *European Energy and Transport, Trends to 2030-Update 2009*. 2010. Available online: http://ec.europa .eu/energy/observatory /trends_2030/doc/trends_to_2030_update_2009.pdf (accessed on 28 July 2011).

30. Moore, G.A. *Crossing the Chasm*; Harperbusiness: New York, NY, USA, 1991.

31. Better Place Website. Available online: http://www. betterplace. com/ (accessed on 2 February 2011).

32. Greening, L.A.; Greene, D.L.; Difiglio, C. Energy efficiency and consumption—The rebound effect—A survey. *Energy Policy* **2000**, *28*, 389–401.

33. Thiel, C.; Perujo, A.; Mercier, A. Cost and CO_2 aspects of future vehicle options in Europe under new energy policy scenarios. *Energy Policy* **2010**, *38*, 7142–7151

Chapter 4

Electric Vehicles – Consumers and Suppliers of the Electric Utility Systems

Cristina Camus[1] and Tiago Farias[1]

[1] Polytechnical Institute of Lisbon - Instituto Superior de Engenharia de Lisboa, Technical University of Lisbon - Instituto Superior Técnico, Portugal

1. INTRODUCTION

Electric vehicles (EVs) have been gaining attention in the last few years due to growing public concerns about urban air pollution and other environmental and resource problems. The technological evolution of the EVs of different types: Hybrid electric vehicles (HEV), battery electric vehicles (BEV) and plug-in hybrid electric vehicles (PHEV), will probably lead to a progressive penetration of EV´s in the transportation sector taking the place of vehicles with internal combustion engines (ICEV). The interesting feature of EVs (only available for BEVs and PHEVs) is the possibility of plugging into a standard electric power outlet so that they can charge batteries with electric energy from the grid.

While a large penetration of plug-in EVs is expected to increase electricity sales, extra generation capacity is not needed if the EVs are recharged at times of low demand, such as overnight hours. EVs, as a local zero emissions' vehicle, could only provide a good opportunity to reduce CO_2 emissions from transport activities if the emissions that might be saved from reducing the consumption of oil wouldn´t be off-set by the additional CO_2 generated by the power sector in providing for the load the EVs represent. Therefore, EVs can only become a viable effective carbon mitigating option if the electricity they use to charge their batteries is generated through low carbon technologies.

In a scenario where a commitment was made to reduce emissions from power generation, the build-up of large amounts of renewable power capacity raises important issues related to the power system operation (Skea, J, et al., 2008), (Halamay et al., 2011), as a result, power system operators need to take measures to balance an increasingly volatile power generation with the demand, and to keep the system reliability. To perform these actions, the SO (system operator) needs to access active and reactive power reserves which are either contractually established with the power generators or traded in the ancillary system market (Estanqueiro, A. et al., 2010). These requirements represent an extra cost for the system which might adequately quantify the negative effect of the variability and uncertainty of each renewable generation technology.

Practically speaking, there are additional external costs of integrating renewable inflexible generation in the power systems, namely in terms of backup capacity, needed to balance power generation and demand when the renewable generation is lower than forecasted, and some kind of storage or demand shift, needed to integrate excesses of renewable generation, especially likely to occur in the off-peak periods.

In this context, electric vehicles can bring techno-economical advantages for the electric power system because of their great load flexibility and increase the system storage capacity. In fact, EVs are parked 93% of their lifetime, making it easy for them to charge either at home, at work, or at parking facilities, hence implying that the time of day in which they charge, can easily vary and, furthermore, for future energy systems, with a high electrification of transportation, Vehicle to Grid (V2G), where the EV works also as an energy supplier, can offer a potential storage capacity and use stored energy in batteries to support the grid in periods of shortage (Kempton and Tomic, 2005). Although each vehicle is small in its impact on the power system, a large number of vehicles can be significant either as an additional charge or a source of distributed generating capacity.

While the aggregate demand for electricity is increasing, decentralized power generation is gaining significance in liberalized electricity markets, and small size electricity consumers are becoming also potential producers. Prosumer is a portmanteau derived by combining the word producer, or provider, with the word consumer. It refers to the evolution of the small size passive consumer towards a more active role in electricity generation and the provision of grid services.

This chapter is concerned with studying how the electric vehicle can work as a "prosumer" (producer and consumer) of electricity. The benefits to the electric utilities and the costs of services provided by EVs in each type of power market will be addressed. The potential impacts of the EVs on the electricity systems, with a great amount of renewable sources in the generation mix will be studied with a focus on the additional power demand and power supply an EV can represent, the role of a new agent on the power market – The EV aggregator – and the economic impacts of EVs on electric utilities.

The analysis of the impact on the electric utilities of large-scale adoption of plug-in electric vehicles as prosumers will be illustrated with a real case study.

Many studies regarding battery electric vehicles and Plug in hybrids have been, and continue to be performed in different countries. In the US, for instance, the capacity of the electric power infrastructure in different regions was studied for the supply of the additional load due to PHEV penetration (Kintner-Meyer et al., 2007) and the economic assessment of the impacts of PHEV adoption on vehicles owners and on electric utilities (Scott et al., 2007). Other studies (Hadley, 2006) considered the scenario of one million PHEVs added to a US sub-region and analyzed the potential changes in demand, impacts on generation adequacy, transmission and distribution and later the same analysis was extended to 13 US regions with the inclusion of GHG estimation for each of the seven scenarios performed for each region (Hadley, 2008). The ability to schedule both charging and very limited discharging of PHEVs could significantly increase power system utilization. The evaluation of the effects of optimal PHEV charging, under the assumption that utilities will indirectly or directly control when charging takes place, providing consumers with the absolute lowest cost of driving energy by using low-cost off-peak electricity, was also studied (Denholm and Short, 2006). This study was based on existing electricity demand and driving patterns, six geographic regions in the United States were evaluated and found that when PHEVs derive 40% of their miles from electricity, no new electric generation capacity was required under optimal dispatch rules for a 50% PHEV penetration. A similar study was made also by NREL (National Renewable Energy Laboratory) but here the analysis focused only one specific region and four scenarios for charging were evaluated in terms of grid impact and also in terms of GHG emissions (Parks et al., 2007). The results showed that off-peak charging would be more efficient in terms of grid stress and energy costs and a significant reduction on CO_2 emissions was expected thought an increase in SO_2 emissions was also expected due to the off peak charging being composed of a large amount of coal generation. Studies made for Portugal (Camus et al., 2011) of the impacts in load profiles, spot electricity prices and emissions of a mass penetration of EV showed that reductions in primary energy consumption, fossil fuels use and CO_2 emissions of up to 3%, 14% and 10% could be achieved by year 2020 in a 2 million EVs' scenario, energy prices could range 0.9€ to 3.2€ per 100 km according to the time of charging (peak and off-peak) and the electricity production mix. A recent report (Grunig M. et al., 2011) that analyzed the EV market for the next years concluded that, the market penetration of EVs will remain fairly low compared to conventional vehicles. The estimation based on several government announcements, industry capacities and proliferation projects sees more than five million new Electric Vehicles on the road globally until 2015 (excluding two- and three-wheelers), the majority of these in the European Union. The main markets for Electric Vehicle are in order of importance the EU, the US and Asia (China and Japan). Some further target markets like Israel and the Indian subcontinent are also expected to evolve. In the long term, the share of EVs will most likely increase as additional countries adopt technologies and initiate projects.

The first description of the key concepts of V2G appeared in 1996, in an article (Kempton and Letendre, 1996) written by researchers at the University of Delaware. In this report the approach was to describe the advantages of peak power to be supplied by EDVs connected to the grid. Further work from the same researchers was continued (Kempton and Letendre, 2002) and the possible power services provided for the grid by vehicles were increased by the analysis of spinning reserve and regulation. The formulation of the business models for V2G and the advantages for a grid that supports a lot of intermittent renewable were described specially for the case of wind power shortage (Kempton and Tomic, 2005a; Kempton and Tomic, 2005b). The use of a fleet for providing regulation down and up was studied and how the V2G power could provide a significant revenue stream that would improve the economics of grid-connected electric-drive vehicles and further encourage their adoption were evaluated (Tomic and Kempton, 2007). The potential impact of renewable generation on the ancillary service market, with a focus on the ability of EVs to provide such services via demand response (DR) and V2G were analyzed. The document also presents a revenue model that incorporates potential scenarios regarding EV adoption, electricity prices, and driver behavior. The output of the model determines the overall revenue opportunity for aggregators who plan to provide DR-EV (Leo M. et al., 2011), although, there is a significantly large market for these services, the limited revenue opportunity for aggregators on a per car basis is unlikely to be compelling enough to justify a business model. According to a recent report from Pike Research (Gibson B., Gartner J., 2011), EVs compete with traditional generation sources as well as emerging technologies, such as stationary battery storage, for revenue from ancillary services such as frequency regulation and demand response.

2. ELECTRICITY GENERATION

Electricity generation faces nowadays a greater number of challenges related to reliability, sustainability and security of supply. The use of renewable resources in power generation has been adopted in most OECD countries as an answer to the climate change problems originated by the burning of fossil fuels in the traditional thermal plants to supply the ongoing increase in electricity demand.

In this section, a description of the electric power systems demand is done emphasizing its evolution along a day and seasonal profile, the different technologies available for power generation are also presented, their main features and when and how each of them produces and the emissions associated with electricity production from thermal units are also addressed in this section. A description of the renewable sources, identifying the factors that influence the value of each renewable technology for the power system is done. These factors include the variability, uncertainty, complementarities with other sources and with the demand and implications for reserve requirements. The impacts of EVs recharge in the typical load profiles will be assessed and also the effects of EVs working as electricity suppliers.

2.1. Electricity Demand And Supply

Electric power systems are designed to respond to instantaneous consumer demand. One of the main features of power consumption is the difference in demand along the day hours, the week days and seasons. This evolution along the day, with a valley during the night that represents about 60% of the peak consumption, has great financial consequences with the need of having several power plants that are useless and an underutilized network during the night.

To supply this load, there are a different set of technologies, from renewable sources (hydro, solar, wind, biomass and waves) to conventional thermal units (natural gas, coal, fuel oil and nuclear).

These different technologies, with different load factors (ratio of average load to capacity), supply the system in different periods and power levels. There are mainly two types of power plants in the electric system: base load or peak power plants. Base load plants are used to meet some of a given region's continuous energy demand, and produce energy at a constant rate, usually at a low cost relative to other production facilities available to the system. Peak power plants are used few hours a year only to fulfill the peaks at higher unit energy prices.

The intermittent renewable sources like the hydro run-of-river and wind are not included in this definition as they are not controllable, but have to be included in the power supply with the highest priority according to the energy and climate policies established (EC, 2009) so they can be considered as base load power plants.

Sometimes the renewable production has an average production profile that works in opposition with demand. Fig. 1 shows as an example, the average production profile of wind power in Portugal verified in year 2010.

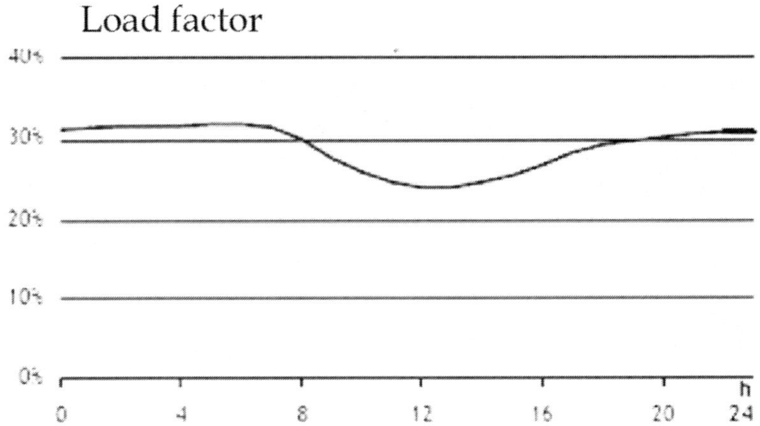

Figure 1. Average wind power profile for year 2010 in Portugal (REN, 2011)

In fact, it has been observed along the years that the wind power production has in average this same profile, with more power production during the night hours.

This situation is even sharper in summer months. In Fig. 2 are the average power produced by wind, solar and small hydro in July 2011 in Portugal.

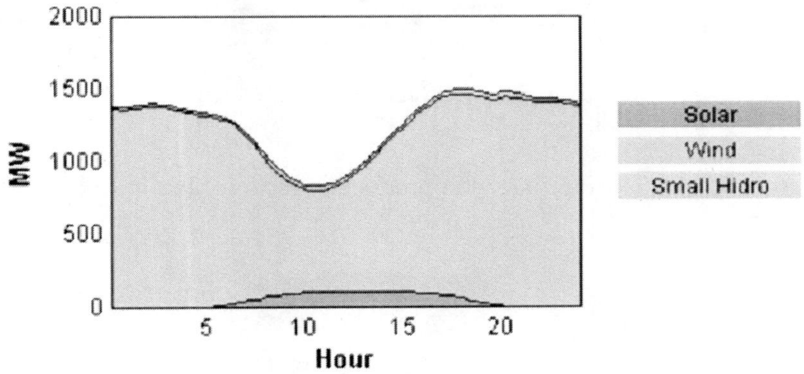

Figure 2. Average power production profile of the renewable sources in July 2011 in Portugal

In summer months, the renewable production is lower when the demand is higher.

For this same case study, Fig. 3 shows the July average power profile with the production technologies. The lowest renewable production level coincides with the peak consumption.

This situation gives the opportunity for electric vehicles contribution for levelling the power consumption diagram and allowing the penetration of more renewable production, by increasing the load during the night hours and supplying the system at the peak hours (Fig. 4).

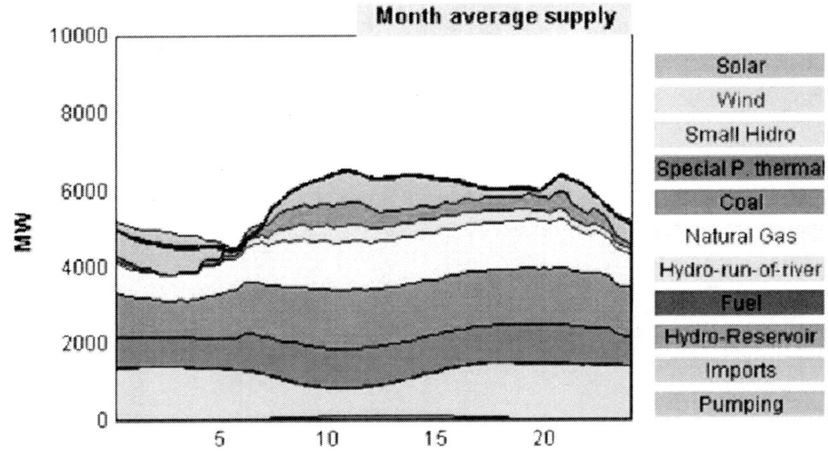

Figure 3. Average load profile with production technologies in July 2011 for Portugal.

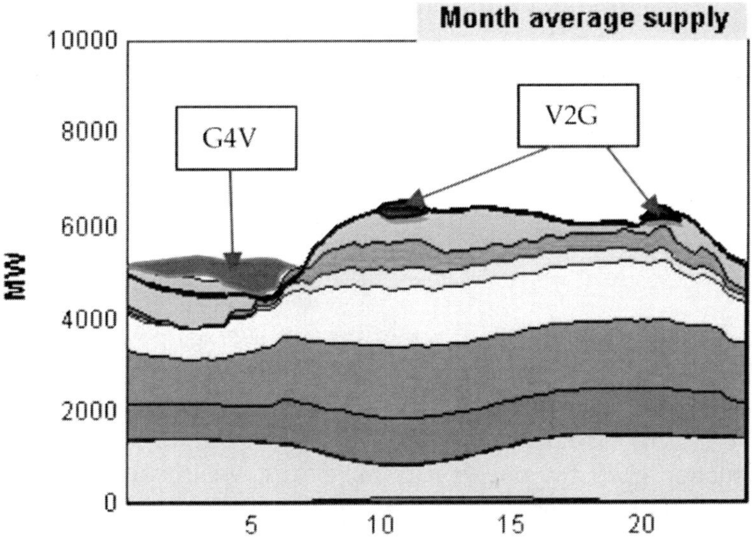

Figure 4. Example of the effect EVs can produce in the electricity demand profile as consumers and suppliers of electricity through G4V (Grid for Vehicle) and V2G respectively.

2.2. The Main Technologies For Electricity Generation And The Merit Order

As described in the previous sub-section, there are many technologies available for electricity production.

The aim of a power plant in a power system is to supply the load in an economical, reliable and environmentally acceptable way. Different power plants can fulfill these requirements in different ways. Different power plants have different characteristics concerning how they can be controlled in the power system. When operating a power system, the total amount of electricity that is provided has to correspond, at each instant, to a varying load from the electricity consumers. To achieve this in a cost-effective way, the power plants are usually scheduled according to marginal operation costs, also known as merit order. Units with low marginal operation costs will operate almost all the time (base load demand), and the power plants with higher marginal operation costs will be scheduled for additional operation during times with higher demand. Wind power plants as well as other variable sources, such as solar and tidal sources, have very low operating costs. They are usually assumed to be zero therefore these power plants are at the top of the merit order. That means that their power is used whenever it is available.

In parallel with marginal operation costs of the power plants are the environmental costs, nowadays assessed by the GHG emissions, they represent. In Table 1, are the average emission rates considered for the typical thermal power plants to compute the GHG emissions from power generation. Those average values can increase if the power plants are subjected to many start-up cycles.

Table 1. Emission rates considered for the thermal power plants for GHG emission computation (EDP, 2008).

Technology	Emission rate (kg/MWh)		
	CO_2	NO_x	SO_2
Coal	900	2.8	6.3
Fuel	830	3.9	4.5
Nat gas (Comb. Cycle)	360	0.13	0
Cogeneration (N.Gas)	600	0.5	

Summarizing, we can dispose of flexible plants, where the power output can be adjusted (within limits), and inflexible plants, where power output cannot be adjusted for technical or commercial reasons. Examples of flexible and inflexible power plants are in Table 2.

As mentioned, the output of the inflexible power plants is treated as given when optimizing the operation of the system.

Not all the flexible power plants can be used the same way to adjust to power demand. The hydro plants with reservoir are the more flexible. Thermal units must be "warmed up" before they can be brought on-line, warming up a unit costs money and start-up cost depends on time unit has been off. There is the need to "balance" start-up costs and running costs. For example a Diesel generator has a low start-up cost but a high running cost, while a Coal plant has a high start-up cost and a low running cost.

Table 2. Available power plants for electricity generation.

Flexible Plants	Inflexible Plants
Coal-fired	Nuclear
Oil-fired	Run-of-the-river hydro
Open cycle gas turbines	Renewable sources (wind, solar,...)
Combined cycle gas turbines	Combined heat and power (CHP, cogeneration)
Hydro plants with storage	

2.3. The Renewable Sources

The percentage of renewable production depends on the location (the endogenous resources available) and the energy policy of the local economy.

Many sources of renewable energy, including solar, wind, and ocean wave, offer significant advantages such as no fuel costs and no emissions from generation. However, in most cases these renewable power sources are variable and non-dispatchable. The utility grid is already able to accommodate the variability of the load and some additional variability introduced by sources such as wind. However, at high penetration levels, the variability of renewable power sources can severely impact the utility reserve requirements.

For instance, at low penetration levels, the variable output of wind power plants is easily absorbed within the variability of the load. However, as the penetration level increases, the added variability of the wind resource can cause greater ramp-rates, greater inter-hour variability, and greater scheduling error. This ultimately increases the amount of generation the system operators must hold in reserve (i.e., the reserve requirement) to accommodate the unplanned excursions in wind generation.

2.3.1. Wind, Solar, And Wave Generation Characteristics

Wind power is now a very mature and established renewable resource throughout the world. However, other renewable power sources such as solar (PV or concentrating/thermal) and ocean wave energy also have significant potential. Each of these renewable power sources can be described by three major characteristics.

1st –Variable. The output power of a large-scale wind, solar, or wave power plant varies over time. The vast majority of the time, the variability from one minute to the next is very small, and even the hourly variation is usually small. However, on occasion the output of a large plant, as high as several hundred MW, may go from full output to low production or vice versa over several hours (Fig. 5);

2nd –Non-dispatchable. As implemented now, the system operator has very limited control of the output of large scale renewable generation. In general, the operator must deal with whatever the renewable generation outputs are in much the same manner as dealing with the load. Therefore it is common in the analysis of the impact of renewable power generation to subtract its contribution from the load: renewable power generation appears as a negative load;

3rd – Energy source. Due to the non-dispatchable nature of wind, solar, or wave, they generally have a relatively low capacity credit. That is, they do not make a significant contribution to the power requirements of the grid for planning purposes. However, each Joule of energy converted by a renewable source is one Joule saved for “traditional” generation, such as coal. Therefore, renewable energy sources can make a significant impact on the energy requirements of the grid.

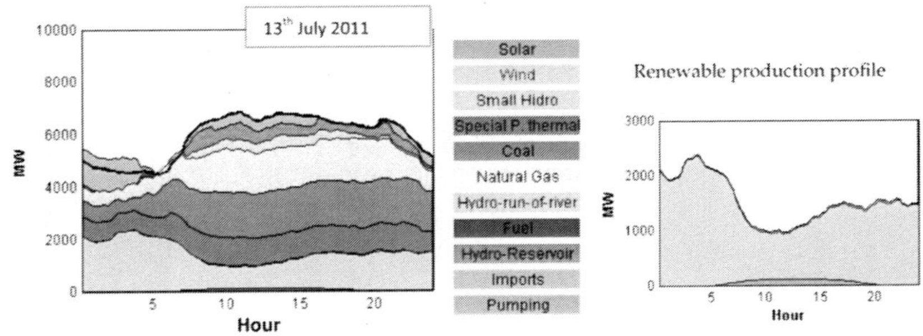

Figure 5. Example of the consumption and production on the 13th July 2011 in Portugal where, in less than 5 hours, a loss of more than 1000MW in renewable production occurred.

2.3.2. Adequacy Of Renewable Production With Power Demand

The variable, non-dispatchable nature of wind, wave, and solar has a significant impact on the utility reserve requirements. Analyzing the effect of these renewable energy sources on the reserve requirements provides a meaningful and concrete method of characterizing the variability of a given renewable energy source, including its short and long-term correlation with the load.

In order to balance generation with load on a minute-by-minute, hourly, or daily basis, the variability of both the generation and the load must be examined.

With renewable resources like wind, solar, and ocean wave, forecasting of the available generation can present a particular challenge, which, while having a large impact on the hourly or daily reserve requirements, often has less of an impact on the intra-hour requirements. Given the focus on reserve requirements, it readily becomes apparent that a clear understanding of the different types/timescales of reserves is necessary.

Three different timescales are currently used to calculate reserve requirements.

The first, regulation, is defined as the difference between the minute-to-minute power generation/load and the 10-minute average power generation/load. This timescale accounts for small changes in power demand or supply that can be readily met through Automatic Generation Control (AGC) via spinning reserves.

The second timescale of interest, following, is defined as the difference between the 10-minute average power generation/load and the hourly average power generation/load.

This timescale accounts for larger changes in the power demand or supply.

The final timescale, imbalance, is defined as the difference between the hourly average power generation/load and the forecasted generation/load for that hour. The imbalance component of the reserve requirements is directly impacted by the accuracy and frequency of the forecasted generation/load. With the large increase in wind power generation, the imbalance component of the reserve requirement is forecasted to grow rapidly.

In order to calculate imbalance reserve requirements, the scheduled or forecasted power must be determined for both the renewable resource and the load.

2.3.3. Energy Storage Needs

Reliability is an important feature of power systems. A reliable power system implies that there is always enough generating capacity to satisfy the power demand. In reality this aim can only be achieved to a certain security level. As the installation of power plants is a long process, future power portfolios and their ability to cover the demand must be assessed in advance. The contribution of wind power to the availability of generating capacity becomes important with increasing wind penetration. The capacity value of wind power is therefore identified for future, potentially large wind power penetration levels.

Capacity value designates the contribution of a power plant to the generation adequacy of the power system. It gives the amount of additional load that can be served in the system at the same reliability level due to the addition of the unit. It is a long established value for conventional power plants. Over recent years similar values have been calculated for wind power. A higher correlation between wind and load will lead to higher capacity values. In the case of low correlation between wind and load, there will be need of more storage capacity to respond to renewable and load in-balances.

The additional requirements and costs of balancing the system on the operational time scale (from several minutes to several hours) are primarily due to the fluctuations in power output generated from wind. A part of the fluctuations is predictable for 2 h to 40 h ahead. The variable production pattern of wind power changes the scheduling of the other production plants and the use of the transmission capacity between regions.

This will cause losses or benefits to the system as a result of the incorporation of wind power. Part of the fluctuation, however, is not predicted or is wrongly predicted. This corresponds to the amount that reserves have to take care of.

The economic, social and political costs of failing to provide adequate capacity to meet demand are so high that utilities have traditionally been reluctant to rely on intermittent resources for capacity. Dimensioning the system for system adequacy usually involves estimations of the LOLP (loss of load probability) index. The risk at system level is the probability (LOLP) times the consequences of the event. For an electricity system, the consequences of a blackout are large, thus the risk is considered substantial even if the probability of the incident is small.

The loss of load expectation (LOLE) is a measure of system adequacy and nominates the expectation of a loss of load event. The required reliability of the system is usually in the order of one larger blackout in 10–50 years.

Since no generating plant is completely reliable, there is always a finite risk of not having enough capacity available. Variable sources may be available at the critical moment when demand is high and many other units fail. Fuel source diversity can also reduce risk.

3. ELECTRICITY MARKET

In terms of the economic model, the electricity industry has evolved from a vertically integrated state-owned monopoly company (not subjected to the normal rules of competition) to a liberalized market where generators and consumers have the opportunity to freely negotiate the purchase and sale of electricity. In this section the typical electricity markets are described and the more adequate markets for EVs are addressed.

3.1. Electricity Market Structure

Electric power systems include power plants, consumers of electric energy and transmission and distribution networks connecting the production and consumption sites. This interconnected system experiences a continuous change in demand and the challenge is to maintain at all times a balance between production and consumption of electric energy. In addition, faults and disturbances should be cleared with the minimum effect possible on the delivery of electric energy.

Power systems comprise a wide variety of generating plant types, which have different capital and operating costs. When operating a power system, the total amount of electricity that is provided has to correspond, at each instant, to a varying load from the electricity consumers. To achieve this in a cost-effective way, the power plants are usually scheduled according to marginal operation costs, also known as merit order. Units with low marginal operation costs will operate almost all the time (base load demand), and the power plants with higher marginal operation costs will be scheduled for additional operation during times with higher demand. Wind power plants as well as other variable sources, such as solar and tidal sources, have very low operating costs. They are usually assumed to be zero therefore these power plants are at the top of the merit order. That means that their power is used whenever it is available.

The electricity markets operate in a similar way, at least in theory. The price the producers bid to the market is slightly higher than their marginal cost, because it is cost-effective for the producers to operate as long as they get a price higher than their marginal costs. Once the market is cleared, the power plants that operate at the lowest bids come first.

If the electricity system fails the consequences are far-reaching and costly. Therefore, power system reliability has to be kept at a very high level. Security of supply has to be maintained both short-term and long-term. This means maintaining both flexibility and reserves that are necessary to keep the system operating under a range of conditions, also in peak load situations. These conditions include power plant outages as well as predictable or uncertain variations in demand and in primary generation resources, including intermittent renewable sources.

3.2. Base Load Power

Base-load power is the "bulk" power generation that is running most of the time. Base-load power is typically sold via long term contracts for steady production at a relatively low price and can better be provided by large power plants because they last longer and cost less per kWh.

3.3. Peak Power

Peak power is used during times of predictable highest demand. Peak power is typically generated by power plants that can be switched on for shorter periods, such as gas turbines and hydro plants with reservoir. Since peak power is typically needed only a few hundred hours per year, it is economically sensible to draw on generators that are low in capital cost, even if each kWh generated is more expensive.

3.4. Spinning Reserve

Spinning reserves are supplied by generators set-up and ready to respond quickly in case of failures (whether equipment failure or failure of a power supplier to meet contract requirements). They would typically be called, say, 20 times per year; a typical duration is 10 min but must be able to last up to 1 h (spinning reserves are the fastest-response and highest-value component of the more general electric market for "operating reserves"). Operation reserves include several types of reserves in place to respond to short-term unscheduled demand fluctuations, or generator/other system failure. Operating reserve represents generators that can be started or ramped up quickly. There are several categories of operating reserves, often referred to as ancillary services.

Quick-start capacity includes combustion turbines and hydroelectricity, while spinning capacity represents other partly loaded fossil and/or hydroelectric plants. The introduction of wind power into a grid can increase these operation-reserve requirements, due to the variability in wind generation.

3.5. Balancing

Balancing or regulation is used to keep the frequency and voltage steady, they are called for only one up to a few minutes at a time, but might be called 400 times per day; Spinning reserves and balancing are paid in part for just being available, a capacity payment per hour available; Base-load and peak are paid only per kWh generated.

The variability in wind generation precludes wind from contributing fully to the reserve margins required by utilities to ensure continuous system reliability.

Planning reserves ensure adequate capacity during all hours of the year. Typical systems require a “peak reserve margin” of 10%-18%. This means a utility must have in place 10%-18% more capacity than their projected peak power demand for the year. This ensures reliability against generator or transmission failure, underestimates of peak demand, or extreme weather events.

Due to the resource variability of wind generation, only a small fraction of a wind farm’s nameplate capacity is usually counted toward the planning reserve margin requirement. In fact, as wind penetrates further into an electric grid, this “capacity credit” for wind generally declines, especially if the wind farms are developed near each other, i.e. if their output is well correlated.

3.6. The Aggregator

To access to the electricity market means, among other aspects, to have access to the so called "market prices". Under this concept, an EV does not have individually the capacity to access to the electricity market, as each quantity of energy produced is insignificant when compared with the regular power players'.

There arises then a new element for the interconnection between the micro-generation and the electricity market, that it can be called by "commercial agent" or "aggregator". The commercial agent or aggregator adds a set of small power producers so that they can became, in a certain way, a fair concurrent in the market by the fact of dealing with a substantial quantity of energy. Under the point of view of the aggregator, there is also the possibility of dealing either with energy generation and/or energy consumption to maximize the economic value of the EV to the consumer and at the same time revenue to the aggregator, it is almost certain that the charging and discharging vehicle will be done in order to allow the vehicle to be charged with the lowest-cost electricity, and also allows the vehicle to provide high-value ancillary services. EVs could be connected to the power system through the aggregator that sells the aggregated demand of many individual vehicles to a utility, regional system operator, or a regional wholesale electricity market. The idea is that EVs respond intelligently to real-time price signals or some other price schedule to buy or sell electricity at the appropriate time so that the vehicles would be effectively "dispatched" to provide the most economical charging and discharging.

4. ELECTRIC VEHICLE AS A CONSUMER AND SUPPLIER OF ELECTRICITY

Given the nature and physical characteristics of EVs, their integration into the grid is performed at the distribution voltage level. Such an interconnection allows each EV to be plugged into the grid to get the energy to charge up the battery. The EVs, when aggregated in sizeable numbers, constitute a new load that the electricity system must supply. However, an EV can be much more than just a simple load given that bi-directional power transfers are possible once the interconnection is implemented. Indeed, the integration allows the deployment of EVs as a generation resource as well as a storage device for certain periods of time when such deployment aids the system operator to maintain reliable operations in a more economic manner. We refer to the aggregated EVs as a generation/storage device in this case. The entire concept of using the EVs as a distributed resource – load and generation/storage device–by their integration into the grid is known as the vehicle-to-grid (V2G). Under this concept, the EVs become active players in grid operations and play an important role in improving the reliability, economics and environmental attributes of system operations. Such benefits include the provision of capacity and energy-based ancillary services, the reduction of the need for peakers and load levelization.

4.1. Electric Vehicle Modeling

Electric vehicles constitute a variety of vehicle types with different battery capacities, vehicle ranges, and vehicle drive trains. Such differences are important to the electric industry because of their influence on daily vehicle electricity consumption. The common characteristic of EVs and PHEVs is that they require a battery, which is the source of all or part of the energy required for propulsion. For EVs, the original energy consumption unit in kWh and the energy consumption per unit distance in kWh/km is generally used to evaluate the vehicle energy consumption. The battery energy capacity is usually measured in kWh and the driving range per battery charge can be easily calculated.

A typical electric vehicle (EV) traction battery system consists of a chain of batteries connected in a series, forming a battery pack with nominal voltages ranging from 72 to 324 V and capable of discharge/charge rates of several hundred amperes.

As vehicles, EVs are not always stationary and, therefore, may be dispersed over a region at any point in time. In a moving state, EVs may be used for commuting purposes or, possibly for longer trips – if the battery capacity is large or if the EV is a PHEV.

For the EVs used for commuting, we can view, therefore, that the vehicles are idle an average of 22 h a day. We note that as the commuting distance is smaller than the potential range of the EVs, not all the energy in the batteries is consumed by the commute. We may see each EV as a potential source of both energy and available capacity that can be harnessed by the grid in addition to supplying the load of the EV to charge up the battery.

In addition to the storage capacity, there are some other aspects of interest in characterizing the batteries. A critically important one is the state of charge (s.o.c.) of the batteries. It is defined as the ratio of the energy stored in a battery to the capacity of the battery. It varies from 0 when the battery is fully discharged to 1– often expressed in percentages as a variation from 0% to 100% – when the battery is fully charged and provides a measure of how much energy is stored in the battery. The s.o.c. typically decreases when energy is withdrawn from the battery and increases when energy is absorbed by the battery. Thus, for a day during which the EV owner goes to work in the morning, parks the EV, goes back home in the late afternoon and then plugs the EV for charging during the night, the s.o.c. will evolve along a pattern illustrated in Fig. 6.

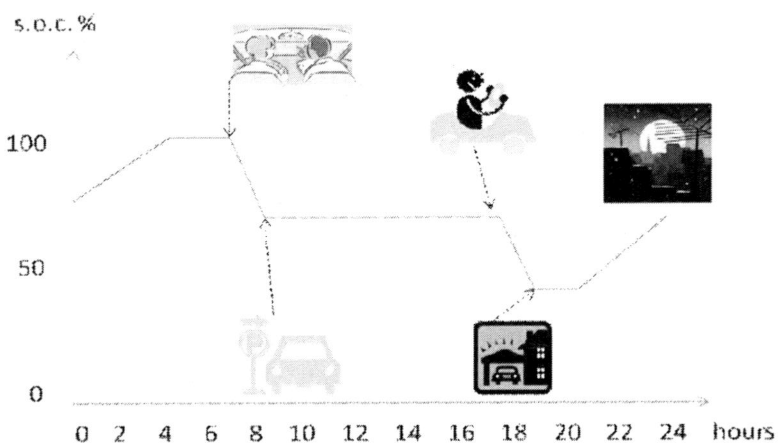

Figure 6. S.o.c. evolution for an EV along a typical working day with only home charging

Batteries release energy more easily when their s.o.c. is high or more exactly above a tolerance level. We stipulate 60% to be the tolerance level in the examples of this work. When the s.o.c. is lower than 60%, a more appropriate utilization of this battery is for energy absorption. If the battery releases energy, then the EV acts as a supply-side resource. If it absorbs energy, the EV acts as a demand-side resource. We can view the battery store present supply-and demand-side resources as a function of the s.o.c. The diagram in Fig. 7 summarizes this information.

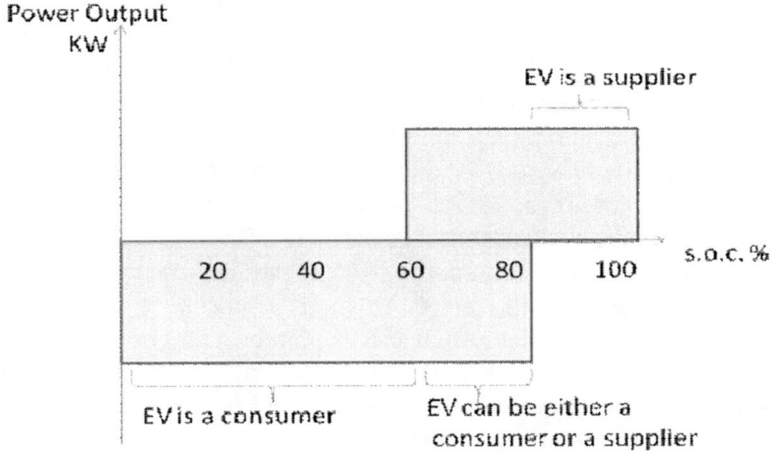

Figure 7. Relation between the s.o.c. and the function of the EV

The frequent switching of the s.o.c. may cause a decrease in battery storage capability which is defined as the battery degradation.

4.2. Evs Aggregation

The battery storage of an individual EV is too small to impact the grid in any meaningful manner. An effective approach to deal with the negligibly small impact of a single EV is to group together a large number of EVs – from thousands to hundreds of thousands. The aggregation, then, can impact the grid both as a load and a generation/storage device.

The basic idea behind such aggregation is the consolidation of the EVs, so that together they represent a load or a resource of a size appropriate to exploit economic efficiencies in electricity markets. The Aggregator is a new player whose role is to collect the EVs by attracting and retaining them so as to result in a MW capacity that can impact beneficially the grid. The size of the aggregation is indeed the key to ensuring its effective role. In terms of load, an aggregation of EVs represents the total capacity of the batteries, an amount in MWs that constitutes a significant size and allows each EV to benefit from the buying power of a large industrial/commercial customer. There are additional economic benefits that accrue as a result of the economies of scale. The aggregated collection behaves as a single decision maker that can undertake transactions with considerably lower transaction costs than would be incurred by the individual EV owners. So, the aggregated entity can make purchases – be it electricity, batteries or other services – more economically than the individual EV owners can and can pass on the savings to each EV owner. As a resource, the aggregated EVs constitute a significant capacity that may beneficially impact the operations of a system operator. The SO deals directly with the Aggregator, who sells the aggregated capacity and energy services that the collection of EVs can provide. The Aggregator's role is to effectively collect the distributed resources into a single entity that can act either as a generation/storage device capable of supplying capacity and energy services needed by the grid or as a controllable load to be connected to the ESP to be charged in a way so as to be the most beneficial to the grid. It is the role of the Aggregator to determine which EVs to select to join the aggregation and to determine the optimal deployment of the aggregation. A single aggregation may function either as a controllable load or as a resource, as depicted in Fig. 8.

The charging of the EVs introduces a new load into the system. For every SO, the load has a typical daily shape formed of on- peak and off-peak periods as described in section 2.

The EV aggregation can act as a very effective resource by helping the operator to supply both capacity and energy services to the grid. To allow the operator to ensure that the supply– demand equilibrium is maintained around the clock, the EV aggregation may be used for frequency regulation to control frequency fluctuations that are caused by supply–demand imbalances. The shape of the regulation requirements varies markedly from the on-peak to the off-peak periods. We define regulation down as the absorption of power and regulation up as the provision of power. A battery may provide regulation up or regulation down service as a function of its s.o.c. Depending on its value for each EV in the aggregation, the collection maybe deployed for either regulation up or regulation down at a point in time. Resources that provide regulation services are paid for the capacity they offer.

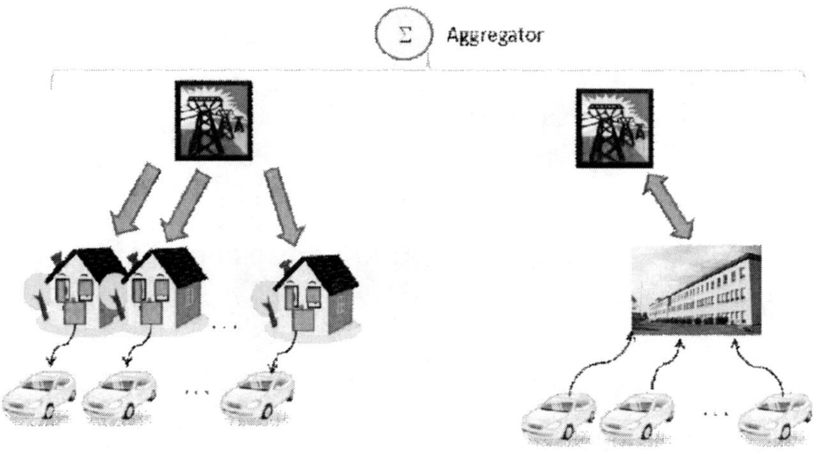

Figure 8. EVs working as load and as supplier of electricity

4.3. The Adequate Electricity Markets For Evs

EVs, with their fast response and low capital costs, appear to be a better match for the quick-response, short-duration, electric services, such as spinning reserves and balancing. The equivalent of those markets in the Portuguese Electric sector, are secondary and tertiary regulation (REN, 2012).

Spinning reserves are paid for by the amount of time they are available and ready even though no energy was actually produced. If the spinning reserve is called, the generator is paid an additional amount for the energy that is actually delivered (e.g., based on the market-clearing price of electricity at that time). The capacity of power available for 1 h has the unit MW-h (meaning 1MW of capacity is available for 1 h) and should not be confused with MWh, an energy unit that means 1MW is flowing for 1 h. These contract arrangements are favorable for EVs, since they are paid as "spinning" for many hours, just for being plugged in, while they incur relatively short periods of generating power.

Regulation or balancing, also referred to as automatic generation control (AGC) or frequency control, is used to fine-tune the frequency and voltage of the grid by matching generation to load demand. Some markets split regulation into two elements: one for the ability to increase power generation from a baseline level, and the other to decrease from a baseline. These are commonly referred to as "regulation up" and "regulation down", respectively. Compared to spinning reserves, it is called far more often, requires faster response, and is required to continue running for shorter durations.

4.4. Estimation Of Costs And Revenues For Vehicles Owners

4.4.1. Estimation Of Revenues For Vehicles Owners

Calculating revenue for vehicle owners depend on the market that V2G power is sold into. Equation 1can be used for markets that pay for available capacity and for energy (Kempton and Tomic, 2005a).

$$r = p_{cap} P t_{plug} + p_{el} R_{d-c} P t_{plug} \tag{1}$$

Where *r* is the total revenue [€], *p cap* is the market price for capacity [€/kW-h], *P* is the contracted capacity available less or equal to *P V2G* [kW], *t plug* is the time the EV is plugged in and available [h],*p el* is the price of electricity for the plugged in hours [cents/kWh], *R d-c* is the dispatch to contract ratio given by *E disp /(P.t plug).*

Capacity payments are an important part of revenue and compensation for energy delivered generally nets out taking into account the energy that must be purchased to charge the vehicle and the cost of batteries depreciation. Furthermore, to compute energy payments, a profile of grid services provided by the vehicle must be defined.

In Portugal, the average capacity prices for regulation between 2007 and 2011 and for the first months of 2012 were shown in Table 3.

In Fig. 9 is depicted the annual average regulation band evolution and the weighed unit capacity price. The average power range for regulation has increased in the last 5 years, representing from 2.8% of average power in 2007 till 5% in 2011.

Table 3. Average prices and capacity for regulation services in Portugal (REN, 2012)

Year	Capacity Price	Power range	Regulation [€/MW]	
	[€/MW-h]	[MW]	up	down
2007	18.5	188	45.7	32.7
2008	21.4	158	63.5	42.2
2009	28.9	197	48.5	21.5
2010	27.2	290	53.8	13.5
2011	27.8	286	73.7	12.6
2012	36.3	291	64,9	24,4

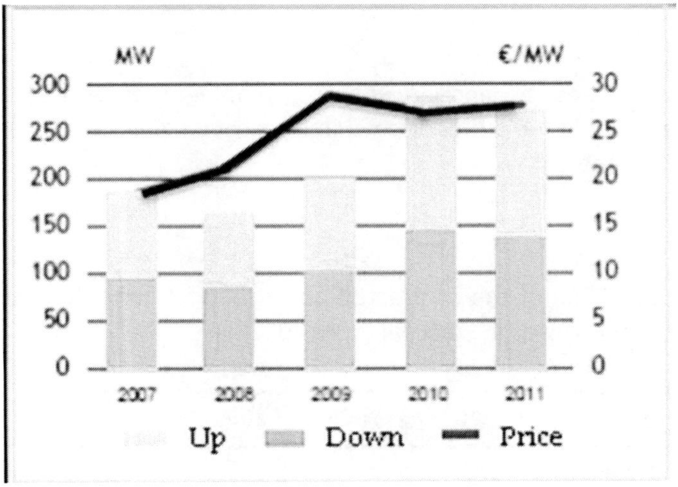

Figure 9. Evolution of regulation band and unit capacity price (REN, 2012)

Looking at the percentage of wind power production in the same 5 years, it increased from 9.3% in 2007 to 18% in 2011. It can be assumed that the increase of intermittent power sources like wind, in the electricity generation mix, leads to an increase of need of power band reserves to assure the same level of system reliability.

In Fig. 10 it is depicted the evolution of capacity installed and energy production in Portugal among the different technologies.

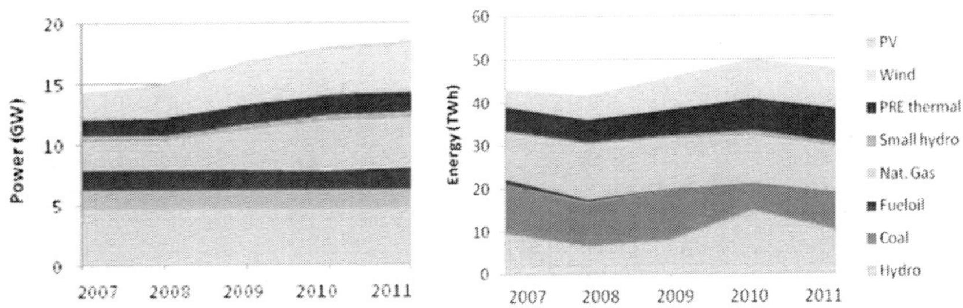

Figure 10. Evolution of capacity installed in the different technologies (left-side) and of annual production from the different technologies (right-side) (REN, 2012)

The increase needs for ancillary services (spinning reserves and regulation) had been fulfilled by the dispatchable technologies in the proportion described in Table 4.

Table 4. Evolution of the contracted power band among the dispatchable available technologies in Portugal (REN, 2012)

Years	2007	2008	2009	2010	2011
Hydro	67%	39%	18%	28%	27%
Coal	10%	3%	16%	12%	14%
Nat.gas	23%	58%	66%	60%	59%

From 2007 till 2011 the power band has increased in 100 MW. To fulfil this 100MW needs, about 30000 EVs at a 3.5 kW each should be plugged. If only

20% of total EVs were available to supply this service, 140000 EVs should be necessary (3% of the total actual light duty fleet). For instance considering the average prices occurred in the first 2012 months and depicted in Table 5.

Table 5. Average prices and capacity for regulation services in Portugal in 2012 (REN, 2012)

Capacity	Power	Regulation [€/MW]			
Price	range	Valley		off-valley	
[€/MW-h]	[MW]	up	down	up	down
36,3	291	64,9	24,4	61,3	35

An EV can expect to achieve a daily revenue of 2.3 € for providing ancillary services to the power grid (Table 6).

Table 6. Expected daily revenues for an EV that provides ancillary services in Portugal

	Capacity	Energy	
		valley	Peak
p_{cap} [cents/kW-h]	3.6		
P [kW]	3.5		
t_{plug} [h]	16	9	7
E_{disp} [kWh/day]		3.0	3.0
P_{elup} [cents/kWh]		6.5	6.1
P_{eldown}[cents/kWh]		2.4	3.5
R_{c-d}		0.10	0.12
Revenue [€/day]	2.02	0.13	0.14

4.4.2. Estimation Of Costs For Vehicles Owners

The cost of V2G is calculated from purchased energy, wear and capital cost. The energy and wear for V2G are those incurred above energy and wear for the primary function of the vehicle, transportation. Similarly, the capital cost is that of additional equipment needed for V2G, but not for driving. The general formula for annual cost is (equation 2):

$$c = c_{en} E_{need} + c_{ac} \tag{2}$$

c is the total cost per year [€], *c en* the cost per energy unit produced for V2G [€/kWh], *E need* is the electric energy needed to be dispatched in the year [kWh] considering the conversion's efficiencies (equation 3).

$$E_{need} = E_{disp} / \eta_{conv} \tag{3}$$

c ac is the annualized capital cost for additional equipment needed for V2G including also the cost of equipment degradation (wear) due to extra use for V2G (equation 4).

$$c_{ca} = c_d + c_c \frac{d}{1-(1+d)^{-n}} \tag{4}$$

c d represents the annual costs of battery degradation, c_c, the capital cost of extra equipment, *d* the discount rate and *n* the investment's life time.

The costs for battery degradation depend on the cycling regimes. As V2G extra cycling would increase battery replacement and additional cost for that should be taken into account. For example considering that a lithium-ion battery could have a 3000 cycle life time (Tomic and Kempton, 2007) at a 100% of discharge and could last almost 10 years with less than a daily charge, an extra shallow, 4% cycling for regulation services occurring in average 10 times in a day it would shorten batteries life in 40% so that after 6 years they should have to be replaced. To compare investments with different life times we use the annuity method (equation 5).

$$c_d = c_{bat} \left(\frac{d}{1-(1+d)^{-n1}} - \frac{d}{1-(1+d)^{-n2}} \right) \tag{5}$$

c bat is the cost of battery and *n 1* and *n 2* are the expected life times without and with V2G.

The estimated costs for EVs' owners for providing ancillary services are depicted in Table 7.

Table 7. Expected annual costs for an EV that provides ancillary services in Portugal

	Costs
Bat Cap [kWh]	16
E_{disp} [kWh]	900
η_{conv}	0,8
E_{need} [kWh]	1125
c_{en} [cents/kWh]	7
c_{bat} [€/kWh]	700
c_d [€/yr]	754
c_c [€]	500
d [%]	8%
n_1 [yr]	10
n_2 [yr]	6
c_{ca} [€/yr]	828
c[€/yr]	907

4.4.3. Estimation Of Financial Results For Vehicles Owners

In this way, estimates for annual profits for EVs' owners, as a result of capacity payments providing regulation capacity could be computed considering the values in table 8.

Table 8. Estimation of costs and revenues for V2G providing regulation services in Portugal

	Revenue	Costs
Capacity [€/yr]	605	
Reg. Up [€/yr]	43	
Reg. down [€/yr]	40	
sav. in recharge [€/yr]	63	
c [€/yr]		907

As V2G is connected at low voltage this regulation service should be purchased by a distribution company that could act as an aggregator to provide enough regulation power to sell in the power markets subjected to the prices shown in Table 5.

We consider that the EVs provide regulation during valley and off-valley hours. During valley hours they are mainly used for charging for further use (for driving and for grid support) but also could provide regulation up and down during this time. For providing regulation up and down we considered the vehicles are plugged-in daily during at least 7 off-valley hours and 9 valley hours. If the vehicles offer this service for 300 days per year a total of 605€ could be earned only for providing. If an average energy of 3.0 kWh is supplied daily to the grid, an annual revenue of 43.2€ could be expected for regulation up and a total revenue of 40€ for regulation down plus 63€ in savings for recharging (due to energy input). Unfortunately, under the described assumptions, total annual costs exceed total revenues in 156€. This loss is very sensitive to battery degradation, if we consider n_2=7 years instead of n_2=6 years, total annual costs decrease to 635€ and a result of 116€ could be obtained (318€ with n_2=8, table 9).

It should not be forgotten that, it is the aggregator that trades directly to the grid for offering regulation services with V2G and works with the market prices showed in table 5 and so a percentage of the earnings should go to the aggregator. Considering a 4% revenue for the aggregator services, the EV's owner profits could range from -156€ to 305€.

Table 9. Estimation of costs and revenues for V2G for different assumptions of battery life

	$n_2=6$	$n_2=7$	$n_2=8$
Total revenue [€/yr]	751	751	751
Total costs [€/yr]	907	635	433
Result [€/yr]	-156	116	318

5. CONCLUSION

Electric vehicles (EVs) and plug-in hybrid electric vehicles (PHEVs), which obtain their fuel from the grid by charging a battery, are set to be introduced into the mass market and expected to contribute to oil consumption reduction. PHEVs and EVs can also provide a good opportunity to reduce CO_2emissions from transport activities if the electricity they use to charge their batteries is generated through low carbon technologies. In addition to the environmental issue, EVs bring techno-economical challenges for utilities as well, because EVs will have great load flexibility as they are parked 93% of their lifetime, making it easy for them to charge either at home, at work, or at parking facilities, hence implying that the time of day in which they charge, can easily vary. EV aggregations can act as controllable loads that contribute to level the off-peak load at night or as generation/storage devices that can provide up and down regulation service when the vehicles are parked.

This chapter described how the electric vehicle can work as a "prosumer" of electricity. The benefits to the electric utilities and the costs of services provided by EVs in each type of power market were addressed, the role of a new agent on the power market – The EV aggregator – and the economic advantages for EVs owners considered the Portuguese energy market as a case study.

There are still many doubts about the life time of EV batteries and battery degradation when proving V2G. Global costs are very sensitive to battery costs and degradation assumptions so that profits can range from -155€/yr to 305€/yr considering respectively 40% to 20% in batteries life range reduction due to V2G supply.

The pressure to generate electricity from endogenous low carbon resources in the majority of the countries makes naturally transport electrification a solution to lower emissions and fossil fuels use from the transportation sector. On the other hand, the increasing of intermittent renewable sources in the power systems, forces the increase of the regulation power band in order to assure the same level of reliability to the power system which would increase the power installed and fixed costs to the power system.

EVs can be a benefit to the environment by reducing emissions and noise in the cities while, at the same time, by providing ancillary services to the power grid, reduce the investments and operation costs in thermal generation and allows the integration of more renewable production. To provide a 100 MW of band power a total of 30000 EVs at a 3.5 kW each should be plugged-in. If only 20% of total EVs are available to supply this service, 140000 EVs should be necessary which corresponds of 3% of the total actual light duty fleet in the Portuguese case study.

ACKNOWLEDGMENT

The authors would like to acknowledge FCT- Fundação para a Ciência e Tecnologia through the national project Power demand estimation and power system impacts resulting of fleet penetration of electric/plug-in vehicles (MIT-Pt/SES-GI/0008/2008).

The authors also thank the REN – Portuguese Energy Networks for supplying up-to-date and valuable data concerning electric power consumption and generation.

REFERENCES

1. Camus C., Farias T., Esteves J., Potential impacts assessment of plug-in electric vehicles on the portuguese energy market, Elsevier Energy Policy 39, Sep. 2011 5883 5897

2. P. Denholm, W. Short, 2006 An Evaluation of Utility System Impacts and Benefits of Optimally Dispatched Plug-In Hybrid Electric VehiclesTechnical Report NREL/TP-620 40293October 2006.10. 2172/ 888683

3. Dr. Duvall, S. Mark, 2006Plug-in hybrid electric Vehicles Technology Challenges", EPRI Sep 2006

4. EC 2009Directive 2009/28/EC of the European Parliament and of the Council of 23 April 2009 on the promotion of the use of energy from renewable sources

5. EDP(2008), Sustainability Report available at http://www.edp.pt/pt/aedp/unidadesdenegocio/producaodeelectricidade/Pages/CompromissoAmbiental.aspx

6. EPRI 2006Comparing the Benefits and Impacts of Hybrid Electric Vehicle Options for compact sedan and sport utility vehicles, EPRI,1006892. July 2002

7. ERSE011 http://www.erse.pt/pt/electricidade/tarifaseprecos/Paginas/default.aspx.last accessed 11/01/15)

8. A. Estanqueiro, C. Mateus, Pestana, R. , Operational " Experience of Extreme Wind Penetrations" in. T. Ackerman, 9th International Workshop on Large-Scale Integration of Wind Power into Power Systems and Transmission Networks for Offshore Wind Power Plants. Québec, CA, 371 383 2010Available at: http://repositorio. lneg. pt/bitstream/10400.9/1172/1/Paper_w159_Estanqueiro.pdf.

9. B. Gibson, J. Gartner, 2011Vehicle to Grid Technologies 2GApplications for Demand Response, Vehicle to Building, Frequency Regulation, and Other Ancillary Services: Market Analysis and Forecasts, Pike Research, 2011

10. M. Grünig, M. Witte, D. Marcellino, J. Selig, H. van Essen, Impact of Electric-Deliverable Vehicles. An overview of Electric Vehicles on the market and in development Delft, CE Delft, April 2011

11. Stanton. W. Hadley, 2006 Impact of Plug-in Hybrid Vehicles on the Electric GridOctober 200610.2172/974613

12. Stanton. W. Hadley, 2008 Potential Impacts of Plug-in Hybrid Vehicles on Regional Power Generation, ORNL/TMJanuary 2008.10.2172/ 932632

13. Kempton W., Letendre S., Electric Vehicles as a new power source for Electric Utilities, Elsevier Science , Vol. 3 157 175Dec 1996

14. Kempton W., Letendre S., The V2G Concept: A New For Model Power? Connecting utility infrastructure and automobiles, Public utilities fortnightly 140(4): 16-26, Feb 2002

15. W. Kempton, J. Tomic, 2005a Vehicle-to-grid power fundamentals: calculating capacity and net revenueJ. Power Sources 144 2005

16. W. Kempton, J. Tomic, 2005bVehicle-to-grid power implementations: from stabilizing the grid to supporting large-scale renewable energy, J. Power Sources 144 2005

17. Kintner-Meyer Michael, Kevin Schneider, Robert Pratt, 2007Impacts assessment of plug-in hybrid vehicles on electric utilities and regional u.s. power grids", Part 1: technical analysis, Pacific Northwest National Laboratory (a), November, 2007.

18. Leo M., Kavi K., Anders H., Moss B, Ancillary Service Revenue Opportunities from Electric Vehicles via Demand Response, School of Natural Resources and Environment University of Michigan, April 2011.

19. K. Parks, P. Denholm, T. Markel, 2007 Costs and Emissions Associated with Plug-in Hybrid Electric Vehicle Charging in the Xcel Energy Colorado Service TerritoryTechnical Report NREL/TP-640 41410May 200710.2172/903293

20. REN, 2012REN- Energy Networks of Portugal- Technical reports, available on line athttp://www.centrodeinformacao.ren.pt/PT/Paginas/ CIHomePage.aspx,last time accessed at 6/03/2012).

21. Michael. J. Scott, Michael. Kintner-Meyer, Douglas. B. Elliott, William. M. Warwick, (2007), Impacts assessment of plug-in hybrid vehicles on electric utilities and regional u.s. power grids: part 2: economic assessment. Pacific Northwest National Laboratory (a), November, 2007.

22. J. Skea, D. Anderson, T. Green, R. Gross, P. Heptonstall, M. Leach, 2008Intermittent renewable generation and maintaining power system reliability", Generation, Transmission & Distribution, IET 2 1January 2008 Page(s):82- 89

23. J. Tomic, W. Kempton, Using fleets of electric-drive vehicles for grid support. J. Power Sources

Chapter 5

Design and Application of a Power Unit to Use Plug-In Electric Vehicles as an Uninterruptible Power Supply

Gorkem Sen [1], Ali Rifat Boynuegri [2], Mehmet Uzunoglu [2,*], Ozan Erdinc [2,3] and João P. S. Catalão [3,4,5]

[1]Department of Electronics and Automation, Ipsala Vocational School, Trakya University, Ipsala, Edirne 22400, Turkey

[2]Department of Electrical Engineering, Faculty of Electric-Electronics, Yildiz Technical University Davutpasa Campus, Esenler, Istanbul 34220, Turkey

[3]Instituto de Engenharia de Sistemas e Computadores—Investigação e Desenvolvimento (INESC-ID), Inst. Super. Tecn., University of Lisbon, Av. Rovisco Pais, 1, Lisbon 1049-001, Portugal

[4]Faculty of Engineering, University of Porto, R. Dr. Roberto Frias, Porto 4200-465, Portugal

[5]Department of Electromechanical Engineering, University of Beira Interior, R. Fonte do Lameiro, Covilhã 6201-001, Portugal

ABSTRACT

Grid-enabled vehicles (GEVs) such as plug-in electric vehicles present environmental and energy sustainability advantages compared to conventional vehicles. GEV runs solely on power generated by its own battery group, which supplies power to its electric motor. This battery group can be charged from external electric sources. Nowadays, the interaction of GEV with the power grid is unidirectional by the charging process. However, GEV can be operated bi-directionally by modifying its power unit. In such operating conditions, GEV can operate as an uninterruptible power supply (UPS) and satisfy a portion or the total energy demand of the consumption center independent from utility grid, which is known as vehicle-to-home (V2H). In this paper, a power unit is developed for GEVs in the laboratory to conduct simulation and experimental studies to test the performance of GEVs as a UPS unit in V2H mode at the time of need. The activation and deactivation of the power unit and islanding protection unit are examined when energy is interrupted.

KEYWORDS

grid-enabled vehicles; distributed generation; plug-in electric vehicles; plug-in hybrid electric vehicles; vehicle-to-home; uninterruptible power supply

1. INTRODUCTION

The world faces an environmental crisis as the volatile prices of petroleum meet rising concerns regarding each nation's energy independence and global warming issues due to greenhouse gas (GHG) emissions. In this regard, the transportation sector plays a crucial and growing role in world energy use and, accordingly, GHG emissions, accounting for approximately 15% of overall GHG emissions [1,2]. These factors contribute to increase interest in alternative vehicle technologies. Nowadays, electrification of transportation has become an important, supported industry trend. Along the past few years, grid-enabled vehicles (GEVs) such as plug-in electric vehicles (PEVs) and plug-in hybrid electric vehicles (PHEVs) are growing in popularity due to increasing governmental regulations on industries and public will to reduce GHG emissions [3]. Given GEVs' popularity, different types are estimated to constitute 35% of the automotive market by 2025 [4]. For this reason, many automotive manufacturers have already started to expand their productions to benefit from the growing GEV market.

In the near future, increasing number of GEVs will be connected to power distribution systems for charging their batteries [5]. In [6], Electric Power Research Institute (EPRI) estimates new vehicle market shares of conventional vehicles (CVs) and PHEVs using a choice-based market modeling of customer choice. Results of this report suggest that CVs will have a market share of 56%, 14%, and 5% and PHEVs will have a market share of 20%, 62%, and 80% by 2050 in low, medium, and high penetration scenarios, respectively. In Morgan Stanley's report [7], specific information is used to forecast sales of GEVs. Market demand of GEV is forecasted to reach one million by 2020. The study of Pacific Northwest National Laboratory (PNNL) [8] examines GEV market penetration scenarios. Yearly market penetration ratios for PHEVs are estimated from 2013 to 2045 for three scenarios: hybrid technology-based assessment, R&D goals achieved, and supply constrained scenarios, in which PHEV market penetration is estimated to reach 9.7%, 9.9%, and 26.9% by 2023, and 11.9%, 29.8%, and 72.7% by 2045, respectively. The study of Oak Ridge National Laboratory (ORNL) [9] estimates that the demand for GEVs will be almost one million by 2015 [9]. It is clear that massive integration of GEVs on the power distribution system will result in a significant increase in electric energy demand and will raise load values at peak times [10]. Also, various studies have already been carried out to predict the effects of GEVs on the power distribution system. As a result of these studies, GEVs have been observed to cause some special effects such as phase imbalance, power quality issues, grid stability, transformer degradation and failure, and circuit breaker and fuse blowout on the power distribution systems [11,12,13,14,15].

All these problems can be prevented with well-designed GEV battery chargers and intelligent charging as part of the smart grid technologies. Massive integration of batteries of GEVs into the power grid can create some opportunities. With smart grid technologies, the function of the GEVs as a mobile energy storage unit in the power grid includes some opportunities such as reactive power compensation, harmonic filtering, voltage support, reducing frequency fluctuations, functioning as an emergency power supply such as an uninterruptible power supply (UPS) which is often named as vehicle-to-home (V2H), improving the effectiveness of home renewable energies by using GEV as storage, load balancing, peak shaving unit [16,17,18,19,20,21,22,23]. However, the design and control of a GEV on-board battery charger is important to perform specified operating conditions.

As it is well known, an energy storage unit that provides the required power for a vehicle is employed in GEVs. This energy storage unit must be recharged at charging points. Nowadays, the interaction of a commercial GEV with the grid is generally unidirectional—the charging process. However, GEVs can be operated bi-directionally by modifying the design of power units of GEV. Thus, the storage units can be supplied to small isolated systems such as single households without the power grid in V2H mode. Due to the large amount of energy stored in batteries of GEVs, they can supply the demand of the consumption center. The general scheme of the V2H operating condition is illustrated in Figure 1. V2H operating conditions would offer the possibility of using a GEV as a domestic back-up power as a UPS [24,25]. Instead, a power unit that can be utilized for the consumption center connection of vehicular systems is important for the realization of this operating condition.

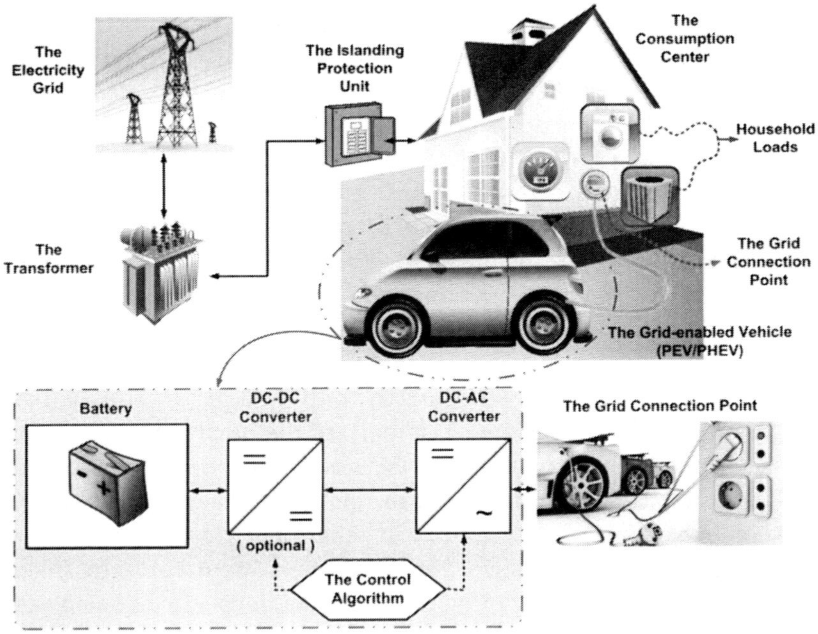

Figure 1. Block scheme of the V2H operating condition.

The main idea presented in this paper is using GEVs as a power supply to the consumption center when needed. The proposed study contributes to the relevant literature regarding the fact that both the simulation and the experimental analyses have been conducted for use of GEVs as UPS. The concept of using GEVs as a UPS unit is different from the regular UPS structures. Firstly, the GEVs are not continuously available at the residential premises as the GEV owners use their GEVs during daily life for traveling to work, *etc.* Besides, the use of GEVs for such purposes should be limited by comfort conditions as the GEV owner requiring a minimum level of charge for his/her GEV for possible unexpected GEV use during the evening, *etc.* Moreover, such systems should be continuously controlled regarding the fact that battery charge/discharge lifetime is limited and uncontrolled daily use of GEVs for such purposes will present drawbacks in this regard. However, the GEVs can be considered as a mobile UPS unit readily available during energy outages, *etc.* for supplying at least the minimum level of electrical energy for sustaining daily activities at home, work, *etc.* as GEVs are also mostly idle during the daytime (at work, *etc.*) even when the GEV is not at home. A limitation for the minimum State-of-Charge (SoC) is implemented within the control structure. In addition, an islanding unit is designed specifically for the mentioned type of use for GEVs in this study.

The simulation studies are performed in MATLAB & Simulink environments. For the experimental studies, a test platform formed by a battery group, a power unit, measurement units, an islanding protection unit, a load unit, a dead-time generation unit, and a dSPACE control unit is configured to test V2H operating conditions—different from the existing literature, which considers the topic a solely theoretical point of view.

The remainder of this paper is organized as follows: Section 2 describes the V2H simulation model that has been developed and explains the physical configuration of the test platform as well as the testing methodology; Section 3demonstrates the simulation and experimental results; and finally, conclusions are made in Section 4.

2. SYSTEM DESCRIPTION AND METHODOLOGY

In order to test and improve the effectiveness of the proposed methodology, an initial analysis is performed in a simulation environment before conducting experimental studies. In this simulation study, the modeling and analysis of the prepared system are realized using MATLAB & Simulink, Sim Power Systems. Afterwards, a test platform is established in order to use GEV as a UPS if necessary. The parameters of the components used in the test platform are normalized to real GEV specifications. The concept of the simulation model and test platform, control algorithms, and the principle of the operation as well as the general structure of the V2H are explained in detail in the following subsections.

2.1. Simulation Model

Firstly, all subsystems of the general structure in the performed simulation study are prepared separately. The simulation study consists of four units: Battery, DC-DC converter, DC-AC converter (inverter), and the load model. The block diagram of the prepared system is shown in Figure 2 and the simulation model is given in Figure 3. Also, the parameter values used in the simulation are given in Table 1. In this simulation study, the charging process of the GEV is not included in order to reduce the complexity in a model that mainly focuses on the V2H operating mode. For this reason, the power grid and the islanding protection unit are not modeled and used in the simulation study. Under normal conditions, islanding operation is important for safety as it separates the power grid and the consumption center. Thus, the electrical energy generated by the GEV is only consumed by the consumption center in the V2H operation mode by this way the staff of the grid operator can safely rectify the fault. Also even when the power grid is powered up again the GEV is temporarily kept mechanically isolated to protect the power converters from transient effects that may occur.

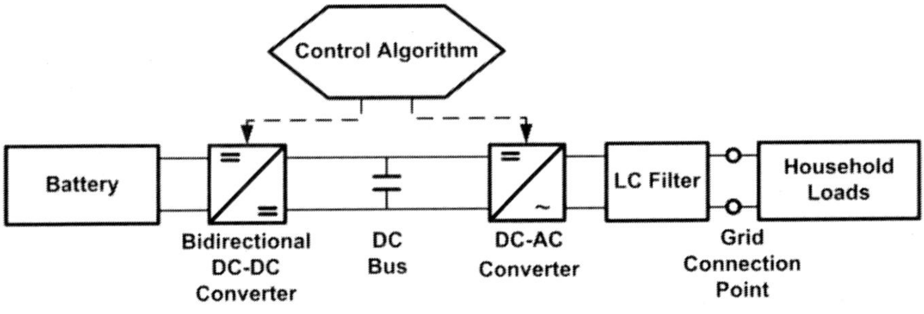

Figure 2. General scheme of the simulation model.

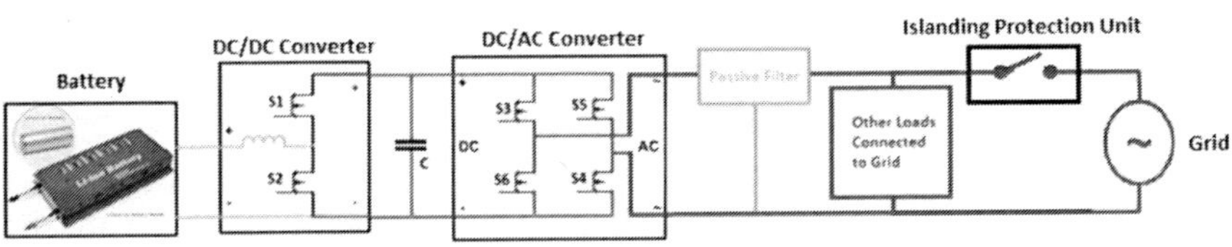

Figure 3. Simulation model.

Table 1. Parameters of the simulation system.

Parameters	Values
Battery Voltage (V)	240
Battery Ah Value (Ah)	10
Battery SoC (%)	100
DC/DC Converter Type	Bidirectional
DC/DC Converter Output Voltage (V)	450
DC/DC Converter Switching Frequency (kHz)	15
DC/AC Converter Type	Full Bridge
DC/AC Converter Control Method	Sinusoidal PWM
DC/AC Converter Switching Frequency (kHz)	10
Filter Inductance Value (mH)	0.03
Filter Capacitor Value (μF)	100

In this study, as the first unit in simulation, the existing battery model within the Simulink library was selected. The type of battery is lithium-ion and the battery values determined as 240 V, 10 Ah, and initial SoC of 100%.

The second unit is a load model that models the consumption center in the simulation study. This load model consists of a controlled voltage source, resistor, and the control block of the controlled voltage source. As the basic operating principle of this system, control block of the load model generates the reference voltage value of the controlled voltage source and current is drawn through 1 ohm virtual resistance. The control inputs of the load unit are active power (P), reactive power (Q), and the instantaneous value of the voltage at the load connection point (V) is used as a feedback signal to calculate reference current value. The mentioned P and Q values of a three-story office building are used. The output of the load model is the voltage signal generated for the controlled voltage source. The signal production method for the controlled voltage source is shown in Figure 4.

First, as in Figure 4, angle (ϕ) is calculated in the unit of radians. The power factor (cos ϕ) value is calculated with ϕ. Also, ϕ is converted to seconds from radians for creating the phase difference between the current and the voltage. Afterwards, the RMS value of the current is calculated so that the peak value of the current can be specified. After all of these processes, a pure sine wave is generated with the same phase angle with the voltage of load connection point. This reference wave form is shifted to ϕ angle and multiplied by the peak value of the current and the reference current signal is produced. Finally, the measured current signal is subtracted from the voltage value on the load connection point. Thus, the control signal for the voltage controlled source is produced.

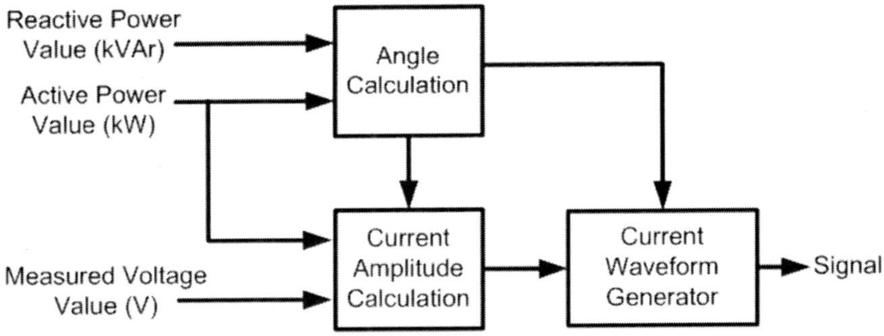

Figure 4. Control block of load model.

The third unit is a DC–DC converter, which generates an appropriate DC voltage for the inverter input. This DC–DC converter requires bi-directional operation (charging and discharging of the battery). In this simulation study, the bi-directional DC–DC converter is operated only for discharging the battery. The charging of the GEV battery is not in the scope of this study; relevant charging methodologies are presented in [24]. Therefore, a bi-directional DC–DC converter control is realized just for the boost operating state. In control unit of the converter, the reference and measured values of DC bus voltage are required. First, the control unit of the DC–DC converter receives the information of the DC bus voltage in the DC–DC converter output. Then, the reference voltage information is compared with the measured voltage in the PI controller. The DC reference voltage value is chosen as 425 V_{DC} for resembling the voltage value at load connection to the grid voltage value. After this process, a pulse width modulation (PWM) signal is obtained by comparison of a saw tooth signal, with the signal received from the output of PI controller. Afterward, zero signal is given to the S1 switch and the PWM signal is sent to the S2 switch. S1 and S2 switches are shown in Figure 3. In order to maintain the DC bus voltage, the control algorithm presented in Figure 5 is used.

The fourth and the last unit is the power unit. This unit includes an inverter and the LC filter, which is located in the inverter output. A sinusoidal pulse width modulation (SPWM) control is applied at the aforementioned inverter. Due to the SPWM method, the switching elements (IGBT, MOSFET) are entered in cutoff mode and transmission mode to generate a sinusoidal wave form during each period. Thus, the variable amplitude sinusoidal signal can be obtained. A sinusoidal control signal is compared with a higher frequency triangle wave for achieving the desired frequency of SPWM. In this comparison result, the SPWM signals required for cross-arms of the power unit are obtained. An LC passive filter is used to filter the harmonics on the output of the inverter. The filter's inductance and capacitor values are selected as 0.03 mH and 100 μF, respectively. The filter is designed and implemented to minimize the system harmonics.

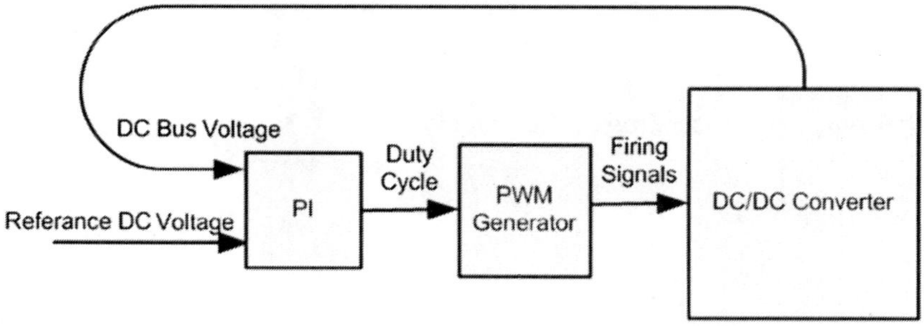

Figure 5. Control algorithm of DC–DC converter during V2H operation mode.

All subsystems in the simulation study are prepared separately, as mentioned before. Then, the interconnection of these subsystems is performed and the simulation results are obtained.

2.2. Test Platform

An experimental test platform is built to verify the controller system design and to implement a successful connection between the GEV and the end-user premise. The parameters of the components used in the test platform are normalized to certain values in order to reduce high costs, minimize the size of the components, and achieve flexible controllability. This test platform consists of a battery, a power unit (inverter), measurement units, an islanding protection unit, a dead-time generation unit, and a dSPACE control unit. The block diagram of the mentioned system is given in Figure 6. The test platform for experimental studies is shown in Figure 7. Also, the values of components used in the test platform are given in Table 2.

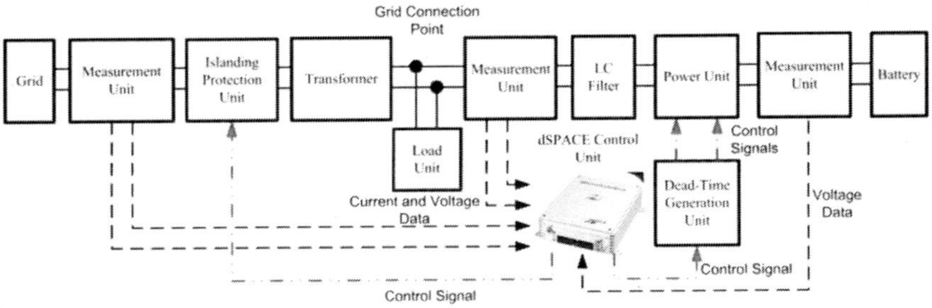

Figure 6. General scheme of the test platform.

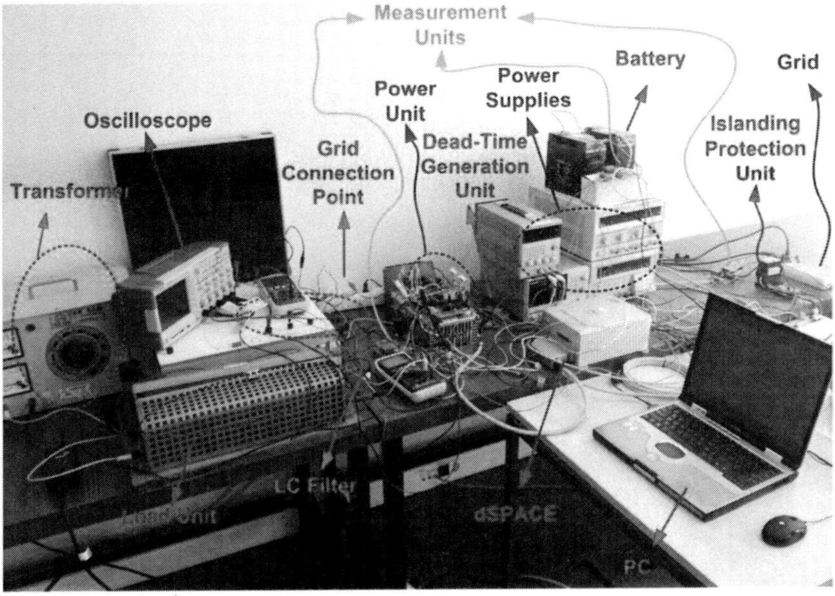

Figure 7. Photograph of the test platform.

Table 2. Parameters of the test platform.

Parameters	Values
Battery Voltage (V)	24
Battery Ah Value (Ah)	18
Battery SoC (%)	100
Power Unit Type	Inverter
Power Unit Control Method	Sinusoidal PWM
Power Unit Switching Frequency (kHz)	10
Filter Inductance Value (mH)	1
Filter Capacitor Value (µF)	15
Load (Ω)	17

2.2.1. Control Algorithm

The control algorithm was developed utilizing MATLAB & Simulink and SimPowerSystems. A dSPACE embedded control unit is used as the controller in experimental studies. The control algorithm of the test platform, as illustrated inFigure 8, includes eight input and two output signals. Five of the input signals are measurement data taken from sensors. At first, these measurement data enter the calibration block and signals are converted to the actual values. Then, the calibrated signals enter the data storage block. The grid voltage information is sent to the grid monitoring block from the data storage block. The information of incoming grid voltage is checked if it is in the voltage range specified by the grid monitoring block. Accordingly, the islanding protection signal is generated. Also, some data are sent to the V2H control block, which provides safe system operation. These data are the battery voltage and the terminal voltage of the power unit, respectively. Furthermore, the V2H control block receives the confirmation signal of the end-user, SoC limit, the price of electricity, and the islanding protection signal from the grid monitoring block. Depending on these inputsV2H operation is started by generating SPWM signal. The SPWM signal generated by dSPACE is applied to the power unit through the dead-time generation unit in order to perform the energy transfer from the battery to the load. In addition to this generated signal, the dSPACE control unit transmits another control signal to the islanding protection unit for commissioning and decommissioning procedures. Lastly, the SPWM block operates according to the confirmation signal of the end-user and generates the driving signal required for the power unit. In this experimental study, the SPWM block for the power unit is operated at 10 kHz.

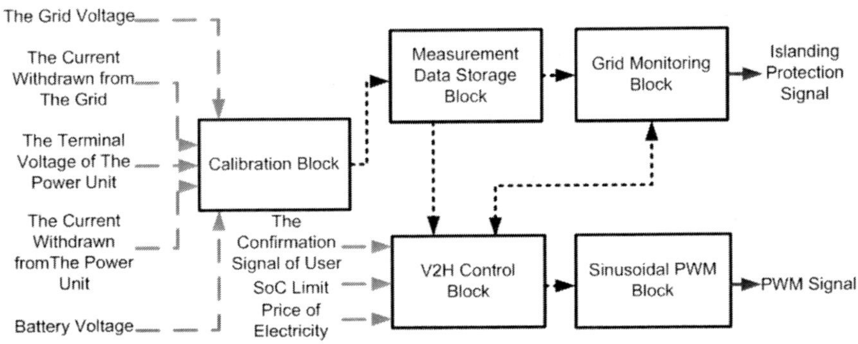

Figure 8. The control algorithm of the test platform.

2.2.2. Operation of the Test Platform

Initially, the battery group and the power unit representing a GEV are connected to the grid connection point via a transformer. However, in this study, GEV is operated as a UPS and therefore the charging of the battery is not included in the experimental study and the power unit is employed only for discharging the battery. The reason for using a transformer in the test platform is the quite low level of battery voltage according to the grid voltage. At the beginning of the experimental study, the grid supplies the load and GEV is in standby mode, depending on the status of the power grid. After a while, the power failure occurs on the grid side and the load does not remain energized. As a result of this fault, the recommended V2H operating mode will take action. From this moment, the power unit requires the formation of the necessary conditions to operate as an UPS. Those conditions are:

- If the initial SoC of the battery is greater than the reference SoC limit,
- If the end-user gives permission for V2H,
- If the islanding protection process is realized (the grid is isolated from loads and power unit),

- If the voltage value of the power unit is approximately zero regarding the measurement unit signal,
- If the price of electricity is greater than the reference price value (optional).

If all of the foregoing conditions are met, the control system formed by dSPACE allows GEV to be operated as a UPS. In this V2H operating mode, the power unit is operated as voltage-controlled and the power flow is provided from the battery to the loads.

The last step is the termination of this operation mode. There are several conditions for the termination of this operation mode. The mentioned conditions for operation termination are:

- If the SoC of the battery is lower than the reference SoC limit,
- If the end-user cancels the operation status,
- If the measurement signals of the grid side show existence of energy (if the voltage value of the power grid is within appropriate values for standards),
- If the price of grid electricity is more than the reference value (optional).

As a result of the provision of one or more termination conditions, dSPACE terminates sending signals to the power unit through the dead-time generation unit. After that, dSPACE receives and evaluates data coming from sensors. As a result of this evaluation, dSPACE interrupts the signal of the islanding protection unit (contactor) when there is no problem in terms of security (approximately two cycles). Consequently, the grid-load connection can be realized again.

3. RESULTS

In this section, the behavior of the GEV in V2H operation mode is examined firstly by the simulation studies (results presented in Section 3.1). After the simulation results are evaluated, experimental results are shown and discussed in Section 3.2.

3.1. Simulation Results

The load demand of a three-story office building located in Istanbul was used in the simulation study. On the day of measurement, building renovation work was in progress. Therefore, some of the electrical loads were continuously engaged or disengaged. In the simulation study, these power values were used as the load demand values for active and reactive power drawn from GEV, as shown in Figure 9. In order to show the simulation results, a period of 10 s was selected when the load demand was at the maximum.

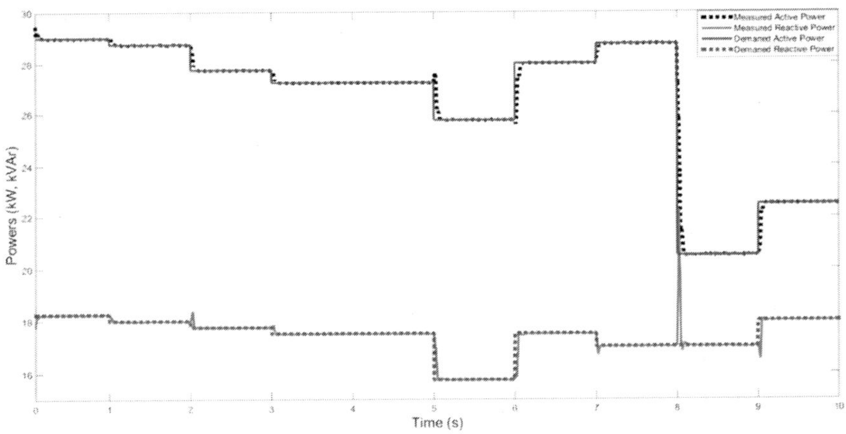

Figure 9. The load demand values and power values drawn from the GEV.

The load demand and the power values drawn from the GEV coincide with each other successfully. An abrupt change was observed at the time of an instant power change due to the instantaneous response of the mathematical operations performed in the simulation study. Eventually, proper functioning of the prepared load model and accuracy of the control of the converters can be observed from Figure 9.

It is necessary to maintain the DC bus voltage at the reference voltage value for maintaining the RMS value of the terminal voltage of the load model the same as the power grid. The mentioned DC bus voltage variation is presented in Figure 10. As seen, some fluctuations can be observed in DC bus voltage as the load power demand varies. During periods of lower power demand, the DC bus voltage can successfully be maintained at the reference voltage value. As the power demand increases, the fluctuations of the DC bus voltage increase as well. However, the controller keeps the mentioned fluctuations within acceptable limits and therefore ensures the successful operation of the system.

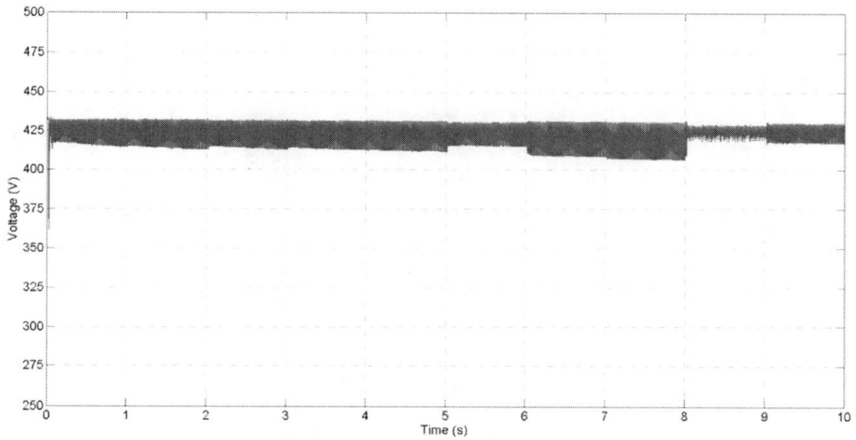

Figure 10. The voltage of the DC bus.

The terminal voltage fluctuations of the power unit are shown in Figure 11. The RMS value of the voltage at the end of load is maintained close to 220 V in spite of high power values drawn from the battery. This voltage fluctuation is within the standards and therefore proper voltage regulation is provided.

Finally, the change of the load voltage value and the current variation drawn by the load in Figure 12 show that the whole system is operating successfully. It is important to notice whether the voltage waveform is kept acceptably steady during load demand variations. The change of load terminal voltage when the highest change of active load occurs in the eighth second, shown in Figure 12, indicates the change of the current drawn by the load at the same time range. After the change of the load values at the eighth second, the prepared system has been successfully adapted to this change and has continued into the current study.

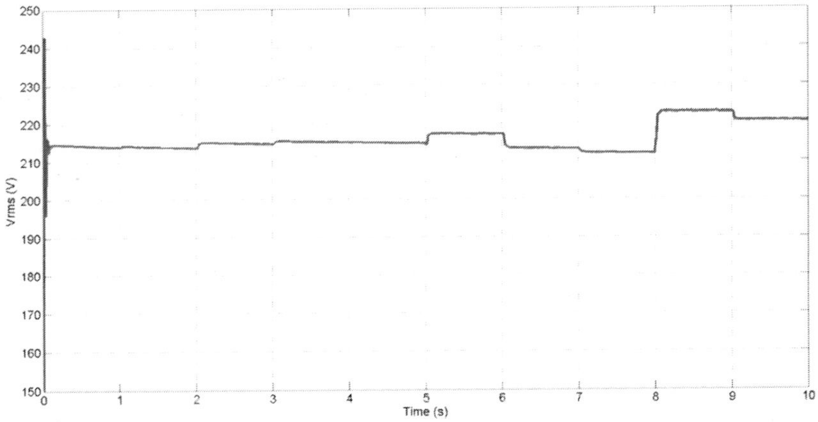

Figure 11. The terminal voltage fluctuations of the power unit.

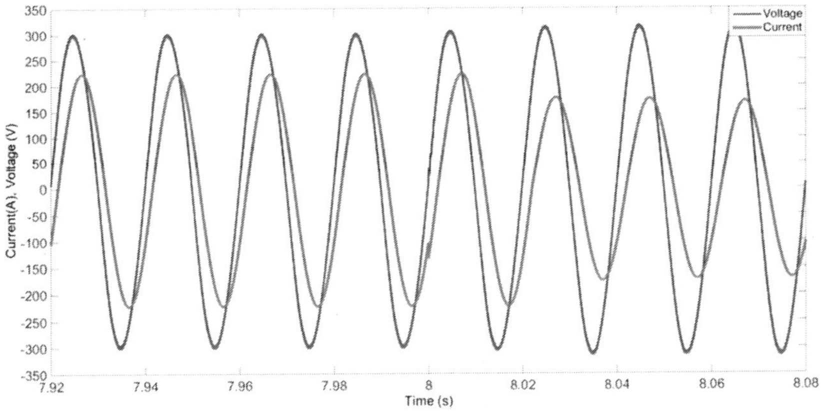

Figure 12. The waveform of the terminal voltage of load model and load current.

Eventually, all units of the simulation system operate successfully, as claimed by the data obtained from all figures. These results show that the GEVs can successfully be used as an UPS if necessary.

3.2. Experimental Results

Once the simulation results validated the control algorithms and V2H operation mode, the test platform was built to support the obtained simulation results. At the beginning of the experimental study, the grid voltage is within specified limits and the grid supplies the load in this process. At the same time, the GEV is awaiting at full charge depending on the grid. After that, the time of the grid voltage is decreasing outside of the specified limits is shown in Figure 13. The dSPACE unit receives the grid voltage and current values from the measurement unit at the grid side and ensures the opening of the contactor through sending a signal to the islanding protection unit. The islanding protection signal, the waveform, and the RMS value of the grid voltage are illustrated in Figure 13.

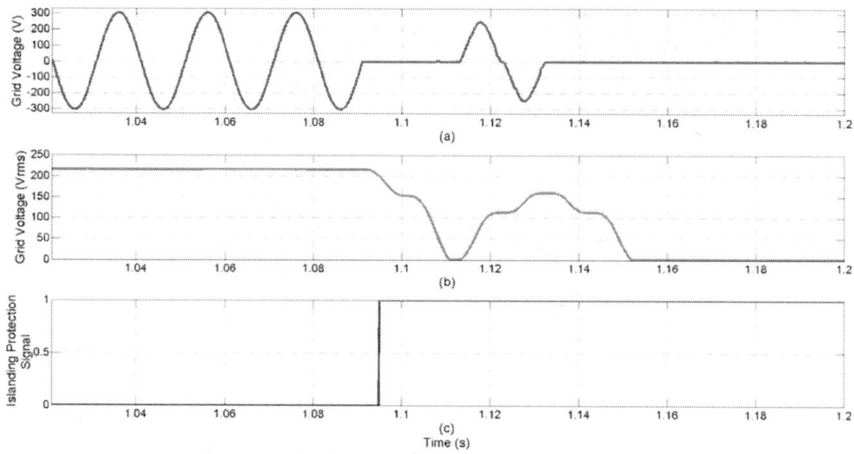

Figure 13. The waveform of the grid voltage when the power grid fails. (**a**) Grid voltage [V] (**b**) Grid voltage [Vrms] (**c**) Islanding protection signal.

The signal shown in Figure 13c is the control signal sent from dSPACE to the islanding protection unit. The islanding protection unit is disabled when the control signal value is 0. Thus, the energy flow from the power grid to the loads is realized. The islanding protection unit is activated and mechanically separates loads and power grid from the moment that the control signal value is 1. Furthermore, the RMS value of grid voltage is reduced to zero after a period of waveforms; this situation is shown in Figure 13b. The main reasons for this situation are that the measuring equipment can measure RMS values after one period and the displayed reduction is within a period of the power grid.

In the first stage of the experimental study, the waveform of the terminal voltage of the power unit is shown in Figure 14a and the RMS value of voltage of the power unit is shown in Figure 14b. After the islanding protection unit is activated, GEV begins to supply loads after a short time. This situation is presented in Figure 14c. Approximately one-period delay is set by dSPACE to secure the operating status of the test platform, which was expressed previously in Section 2.

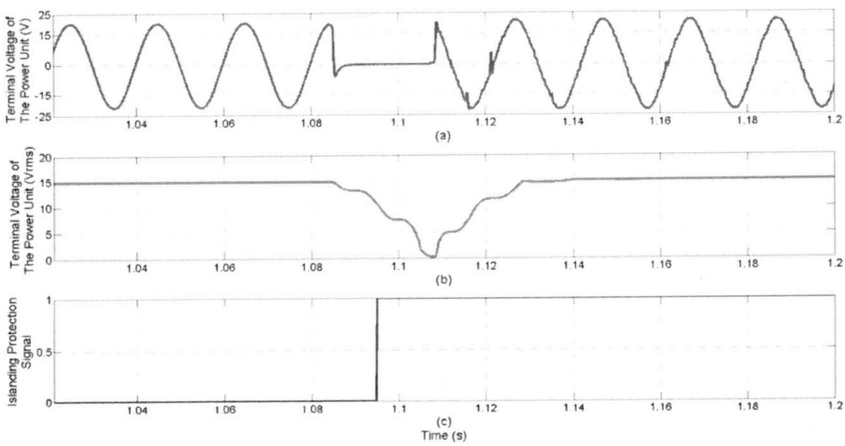

Figure 14. The waveform of the terminal voltage of the power unit when the power grid fails. (**a**) Terminal voltage of the power unit [V] (**b**) Terminal voltage of the power unit [Vrms] (**c**) Islanding protection signal.

Lastly, the change of the battery voltage is shown in Figure 15 at the time of the islanding protection signal received. Initially, the fully charged battery is in standby mode until the islanding protection process is performed. After the islanding protection process, the battery voltage remains at the same value while the inverter is waiting for the provision of the necessary conditions for operation. When the power unit starts, the battery supplies the load and the battery voltage value decreases slightly.

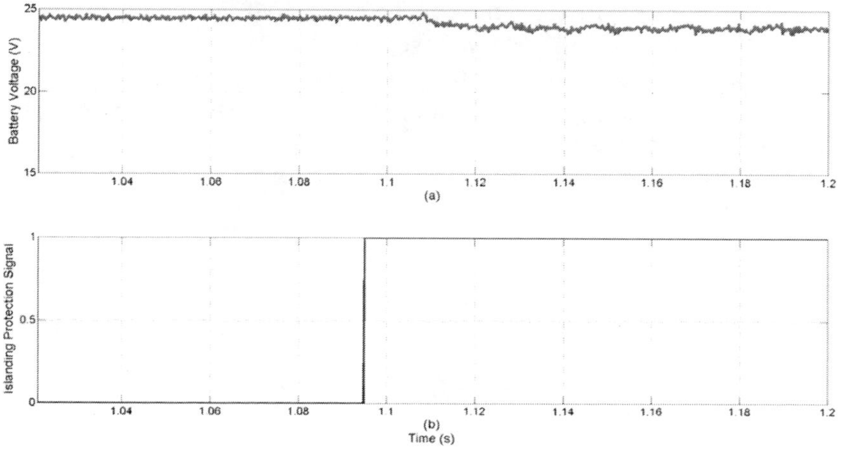

Figure 15. The change of the battery voltage when the power grid fails. (**a**) Battery voltage [V] (**b**) Islanding protection signal.

In the last stage of the experimental study, the response of the test platform is examined when the power grid is powered up again. When the voltage value of the grid is within standard limits again, the flow of energy from GEV to the load is terminated and load demand is satisfied by the grid again. These situations are illustrated in Figure 16, Figure 17 and Figure 18.

The waveform of the grid voltage and its RMS value are shown in Figure 16a,b respectively, when the power grid is online again. As seen in Figure 16c, the value of the islanding protection signal is 0 shortly after the grid voltage is within specified limits. Thus, the islanding protection unit remains inactivated and the grid supplies the loads. Despite the improvement of the grid voltage, the value of the islanding protection signal is zero again after a certain time. The main reason for this is that dSPACE waits for the realization of the previously mentioned security requirements in Section 2.2.2. These safety requirements ensure the safe and effective operation of the whole system.

Figure 17 shows the change of the terminal voltage of the power unit and the islanding protection signal when the power grid is online again. After receiving the information indicating that the grid voltage is within specified limits, the operation of the power unit is terminated. Before islanding operation is terminated, the RMS value of the output voltage of the power unit reaches nearly zero in terms of security. Thus, the grid can be used to supply the load again. Lastly, the change of the battery voltage is shown in Figure 18. Primarily, the grid voltage reaches the specified limit. Then, dSPACE stops the operation of the power unit. From this moment on, the battery voltage is restored to the initial values. After obtaining all the safety requirements, dSPACE terminates the islanding protection.

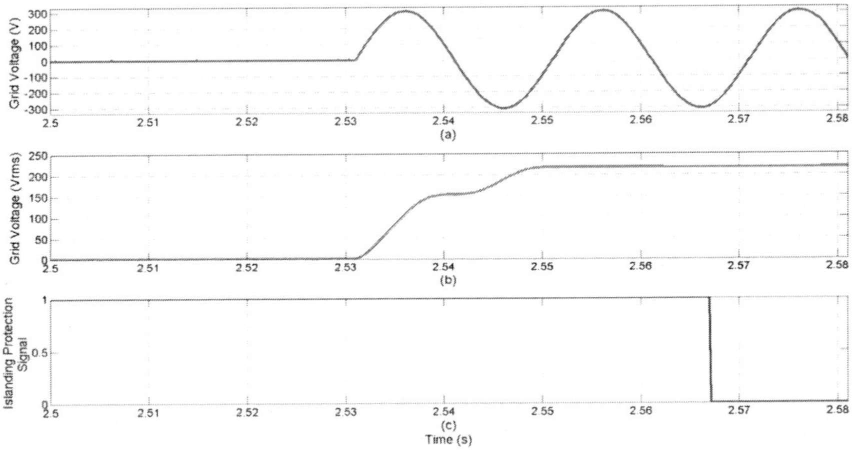

Figure 16. The waveform of the grid voltage when the power grid is powered up again. (**a**) Grid voltage [V] (**b**) Grid voltage [Vrms] (**c**) Islanding protection signal.

The results of the experimental study show that all the units of the test platform perform successfully under the normalized conditions. Consequently, GEVs can be used as a UPS if the electrical infrastructure and power unit control are prepared properly.

4. CONCLUSIONS

Recently, a great number of studies have been performed on GEVs, which are considered to provide zero emissions. In these studies, it is declared that this GEV technology will gradually become widespread in the near future. As it is well known, an energy storage unit that provides the required power for a vehicle is employed in GEVs. This energy storage unit should be recharged at charging points. Also, another reason for this type of GEV to be popular is that the vehicle can be connected to the grid. Nowadays, the interaction of GEV with the grid is bi-directional, as is the charging process. However, GEVs can be operated bi-directionally by modifying the design of a vehicle's power unit or units. In such operating condition, a GEV can be operated as a UPS and supply the energy demand of the consumption center.

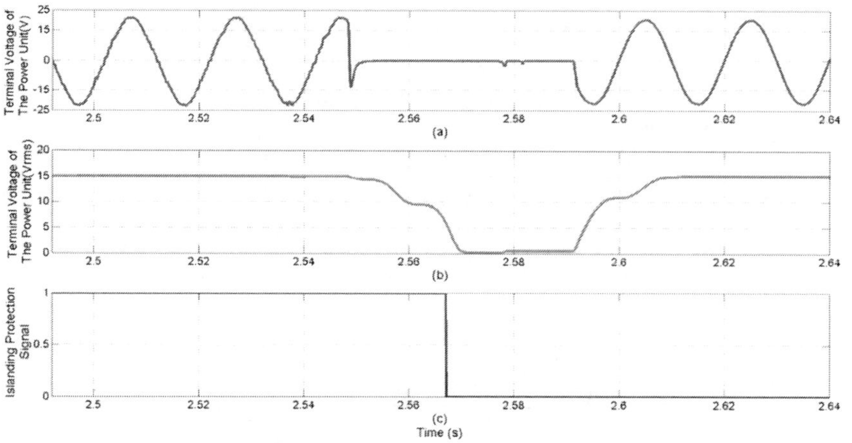

Figure 17. The waveform of the terminal voltage of the power unit when the power grid is powered up again. (**a**) Terminal voltage of the power unit [V] (**b**) Terminal voltage of the power unit [Vrms] (**c**) Islanding protection signal.

Before starting the study, it was assumed that battery of a GEV can supply a portion of an end-user premise's load. The test system was prepared primarily in a simulation environment for testing and improving the accuracy of the planned system and algorithm and the system was tested with the real measured power data of a building. The voltage, frequency, *etc.*, values remained within standards and the building could be supplied with high-quality energy. The results of the simulation study showed that the proposed test system can be used successfully in daily life as a UPS.

Afterwards, the test platform was established to present the verification of the controller system design and to implement a successful vehicle–home connection. The experimental results showed that the established test platform performed successfully under normalized conditions and can be practically implemented. The activation and deactivation of the power unit and islanding protection unit were realized in as rapid (approximately two cycles) and safe a way as possible when energy was interrupted on the test platform.

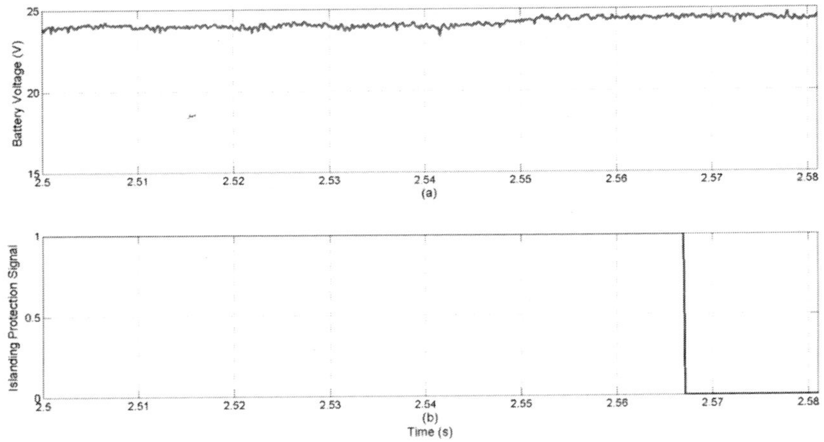

Figure 18. Change of the battery voltage when the power grid is powered up again. (**a**) Battery voltage [V] (**b**) Islanding protection signal.

To this end, it should be noted that the idea of the V2H operating conditions has been confirmed by successful simulation and experimental studies. It is expected that the developed test platform can also be adapted to any GEV and end-user type, thus enabling GEVs to be employed for other purposes apart from transportation. In addition, this concept can also be used in different situations, such as emergency disasters such as earthquakes and blackouts.

ACKNOWLEDGMENTS

This work was supported in part by the Istanbul Chamber of Industry research projects fund. Besides, this work was supported by FEDER funds through COMPETE and by Portuguese funds through FCT, under FCOMP-01-0124-FEDER-020282 (Ref. PTDC/EEA-EEL/118519/2010), UID/CEC/50021/2013 and SFRH/BPD/103744/2014. Also, the research leading to these results has received funding from the EU Seventh Framework Programme FP7/2007-2013 under grant agreement no. 309048 (project SiNGULAR).

AUTHOR CONTRIBUTIONS

Gorkem Sen and Ali Rifat Boynuegri mainly contributed by realizing the simulation and experimental analyses. Besides, all co-authors contributed by significant technical suggestions and improvements through the research steps and also contributed to the writing of the final research paper.

REFERENCES

1. Plug-in Hybrid Electric Vehicles. Available online: http://ns.umich.edu/Releases/2009/Oct09/PHEV_Curtin.pdf (accessed on 7 January 2016).

2. Reducing Transport Greenhouse Gas Emissions: Trends & Data 2010. Available online: http://www.internationaltransportforum. org/Pub/pdf/10GHGTrends.pdf (accessed on 7 January 2016).

3. Tarroja, B.; Eichman, J.D.; Zhang, L.; Brown, T.M.; Samuelsen, S. The effectiveness of plug-in hybrid electric vehicles and renewable power in support of holistic environmental goals: Part 1—Evaluation of aggregate energy and greenhouse gas performance. *J. Power Sources* **2014**, *257*, 461–470.

4. Harrop, P.; Das, R. Hybrid and Pure Electric Cars 2009–2019. Available online: http://www.idtechex.com (accessed on 7 January 2016).

5. Haidar, A.M.A.; Muttaqi, K.M.; Sutanto, D. Technical challenges for electric power industries due to grid-integrated electric vehicles in low voltage distributions: A review. *Energy Convers. Manag.* **2014**, *86*, 689–700.

6. Duvall, M.; Knipping, E. *Environmental Assessment of Plug-in Hybrid Electric Vehicles, Volume 1: Nationwide Greenhouse Gas Emissions*; Technical Report; Electric Power Research Institute (EPRI): Palo Alto, CA, USA, 2007.

7. Stanley, M. *Plug-in Hybrids: The Next Automotive Revolution*; Technical Report; Morgan Stanley: New York, NY, USA, 2008.

8. Balducci, P. *Plug-in Hybrid Electric Vehicle Market Penetration Scenarios*; Technical Report; U.S. Department of Energy: Washington, DC, USA, 2008.

9. Sikes, K.; Gross, T.; Lin, Z.; Sullivan, J.; Cleary, T.; Ward, J. *Plug-in Hybrid Electric Vehicle Market Introduction Study: Final Report*; Technical Report; Oak Ridge National Laboratory (ORNL): Washington, DC, USA, 2010.

10. Rei, R.J.; Soares, F.J.; Rocha Almeida, P.M.; Peças Lopes, J.A. Grid interactive charging control for plug-in electric vehicles. In Proceedings of the 13th International IEEE Annual Conference on Intelligent Transportation Systems, Madeira Island, Portugal, 19–22 September 2010.

11. Liu, R.; Dow, L.; Liu, E. A survey of PEV impacts on electric utilities. In Proceedings of the IEEE PES Innovative Smart Grid Technologies Conference, Anaheim, CA, USA, 17–19 January 2011.

12. Putrus, G.A.; Suwanapingkarl, P.; Johnston, D.; Bentley, E.C.; Narayana, M. Impact of electric vehicles on power distribution network. In Proceedings of the IEEE Vehicle Power and Propulsion Conference, Dearborn, MI, USA, 7–10 September 2009.

13. Farmer, C.; Hines, P.; Dowds, J.; Blumsack, S. Modeling the impact of increasing PHEV loads on the distribution infrastructure. In Proceedings of the 43rd Hawaii International Conference, Honolulu, HI, USA, 5–8 January 2010.

14. Roe, C.A. *Power System Impacts of Plug-in Hybrıd Electrıc Vehicles*; Georgia Institute of Technology: Atlanta, GA, USA, 2009.

15. Maitra, A.; Taylor, J.; Brooks, D.; Alexander, M.; Duvall, M. Integrating plug-in electric vehicles with the distribution system. In Proceedings of

the CIRED 20th International Conference on Electricity Distribution, Prague, Czech Republic, 8–11 June 2009.

16. Kisacikoglu, M.C.; Ozpineci, B.; Tolbert, L.M. Effects of V2G reactive power compensation on the component selection in an EV or PHEV bidirectional charger. In Proceedings of the Energy Conversion Congress and Exposition (ECCE), Atlanta, GA, USA, 12–16 September 2010.

17. Kempton, W.; Udo, V.; Huber, K.; Komara, K.; Letendre, S.; Baker, S.; Brunner, D.; Pearre, N. *A Test of Vehicle-to-Grid (V2G) for Energy Storage and Frequency Regulation in the PJM System*; University of Delaware: Newark, DE, USA, 2009.

18. Wang, X.; Tian, W.; He, J.; Huang, M.; Jiang, J.; Han, H. The application of electric vehicles as mobile distributed energy storage units in smart grid. In Proceedings of the Power and Energy Engineering Conference (APPEEC), Wuhan, China, 25–28 March 2011.

19. Masoum, M.A.S.; Moses, P.S.; Deilami, S. Load management in smart grids considering harmonic distortion and transformer derating. In Proceedings of the International Conference on Innovative Smart Grid Technologies (ISGT), Gaithersburg, MD, USA, 19–21 January 2010.

20. Tascikaraoglu, A.; Boynuegri, A.R.; Uzunoglu, M. A demand side management strategy based on forecasting of residential renewable sources: A smart home system in Turkey. *Energy Build.* **2014**, *80*, 309–320.

21. Bizon, N. Energy efficiency for the multiport power converters architectures of series and parallel hybrid power source type used in plug-in/V2G fuel cell vehicles. *Appl. Energy* **2013**, *102*, 726–734.

22. Habib, S.; Kamran, M.; Rashid, U. Impact analysis of vehicle-to-grid technology and charging strategies of electric vehicles on distribution networks—A review. *J. Power Sources* **2014**, *277*, 205–214.

23. Babrowski, S.; Heinrichs, H.; Jochem, P.; Fichtner, W. Load shift potential of electric vehicles in Europe. *J. Power Sources* **2014**, *255*, 283–293.

24. Boynuegri, A.R.; Uzunoglu, M.; Erdinc, O.; Gokalp, E. A new perspective in grid connection of electric vehicles: Different operating modes for

elimination of energy quality problems. *Appl. Energy* **2014**, *132*, 435–451.

25. Tuttle, D.P.; Fares, R.L.; Baldick, R.; Webber, M.E. Plug-in vehicle to home (V2H) duration and power output capability. In Proceedings of the Transportation Electrification Conference and Expo (ITEC), Detroit, MI, USA, 16–19 June 2013.

Chapter 6

Electric vehicle traction motors without rare earth magnets

James D. Widmer, Richard Martin, Mohammed Kimiabeigi

Centre for Advanced Electrical Drives, School of Electrical and Electronic Engineering, Newcastle University, Newcastle-upon-Tyne, UK

ABSTRACT

Since the Hybrid Electric Vehicle (HEV) became main-stream with the launch of the Toyota Prius in 1997, the use of rare earth magnets in vehicle traction motors has become common. In particular the rare earth based, hard magnetic material Neodymium Iron Boron (NdFeB) has offered significant performance benefits, not possible with other technologies, enabling the development of compact, torque- and power-dense electric traction motors. This trend has continued as mass market Battery Electric Vehicles (BEV) such as the Nissan Leaf, have come to market. However in 2011–2012 the price of these materials rose significantly, owing to geopolitical concerns relating to security of supply. Whilst the price has recovered more recently closer to historical levels, concern still remains in the minds of governments and many manufacturers of hybrid and electric vehicles. Reports have also raised questions over the environmental sustainability of these materials and this has further encouraged users to consider alternatives. This paper therefore examines why these magnetic materials have been so successful in traction motor applications. It also explores the alternatives, including those which are ready for market and those which are in the process of being developed.

KEYWORDS

Electrical machines; Rare earth magnets; Electric vehicles; Traction motors

1. INTRODUCTION

1.1. Neodymium Iron Boron magnets and electric traction motors

The sintered Neodymium Iron Boron (NdFeB) hard magnetic material was patented by Sumitomo Special Metals in 1983 [1] (Fig. 1) and brought about a step change in terms of electric motor performance. Neodymium is a member of the family of materials known as Light Rare Earth Elements (LREE), along with others including for example Lanthanum (used in optics) and Samarium (also used in magnetic materials).

Figure 1. (Left) An example of a sintered NdFeB magnet and (right) a vehicle traction motor, the 80 kW interior permanent magnet motor from the Nissan Leaf, in which such magnets are used.

These magnets offer such high levels of performance owing to their very high Maximum Energy Product compared to other magnetic materials (Fig.

2). The Maximum Energy Product is the measure of the magnetic energy which can be stored, per unit volume, by a magnetic material; it is calculated as the maximum product of a material's residual magnetic flux density (degree of magnetisation) and its coercivity (the ability to resist demagnetisation once magnetised). Equally, if each of these constituent properties, the remnant flux density and coercivity, is taken separately then in NdFeB they are in their own right significantly higher than for other magnetic materials (Fig. 3) [2], [3] and [4].

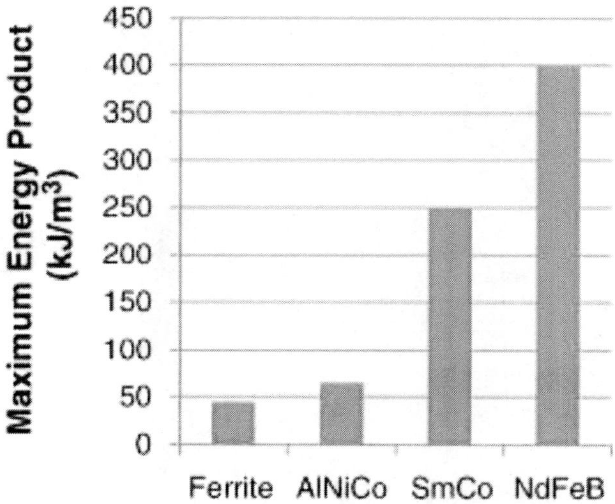

Figure 2. Comparison between the Maximum Energy Product of differing hard magnetic materials.

It is also important to recognise that a key ingredient in allowing NdFeB magnets to operate at high ambient temperatures is Dysprosium; this Heavy Rare Earth Element (HRE) is added to NdFeB in order to increase the high temperature coercivity (ability to withstand demagnetisation) of the magnets above circa 100 °C [5]. This has been essential in making it possible to use these magnets in high power density applications, such as vehicle traction.

In an electric traction motor, NdFeB magnets allow a very strong magnetic field to be generated in a very small volume. The alternative would be to use electromagnets, where a magnetic field is generated by passing current through a conducting coil. It can be shown that a 3 mm thick piece of NdFeB magnet produces the equivalent magnetic field to passing 13 A (being the rating of a UK home electrical socket) through a coil with 220 turns of copper wire. In terms of space, if a current density of 10 A/mm^2 is assumed in the conductor (which is typical for normal operation of a traction motor), then an equivalent electromagnetic coil might have five times the cross sectional area of the NdFeB magnet (Fig. 4). At the same time the coil would produce losses in the windings of 50 W or more per metre length of the coil, arising due to the electrical resistance of the conductor. To put this in perspective, in a representative 80 kW traction motor the optimum use of NdFeB magnets would theoretically be equivalent to saving perhaps 20% in total motor volume and, conservatively, in the order of 300 W of winding loss.

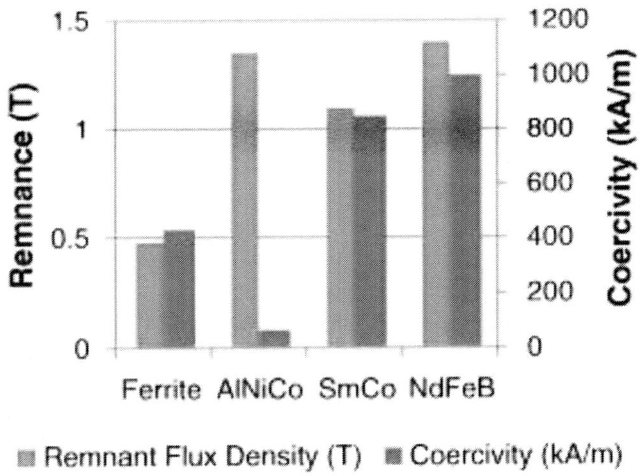

Figure 3. Comparing remnant flux density (magnetic strength) and coercivity (resistance to demagnetisation) for different hard magnetic materials.

1.2. Concerns with rare earth magnets

Although the performance benefits are undisputed, the use of rare earth materials, such as NdFeB, has raised concerns in a number of areas. In 2011 and 2012, China reportedly threatened to cut off international supplies of these materials [6]. Concern over the availability of these materials then led to dramatic, though short-term, increases in the price of many rare earth materials, with Neodymium and Dysprosium prices both increasing by an order of magnitude (Fig. 5). Prices have since recovered to nearer their early 2011 levels, however in the case of Dysprosium, they have not yet returned to their previously low levels. It is reported that as a result of this price spike, a number of organisations are seeking to open new mines. However these have focused on light rare earth oxide extraction (for example, Neodymium) [7]. It appears that heavy rare earth oxide mining remains focussed on extraction from ionic absorption clays found in southern China. This suggests that light rare earths (Neodymium) will remain at relatively low costs whilst heavy rare earths (Dysprosium) may remain in relatively short supply.

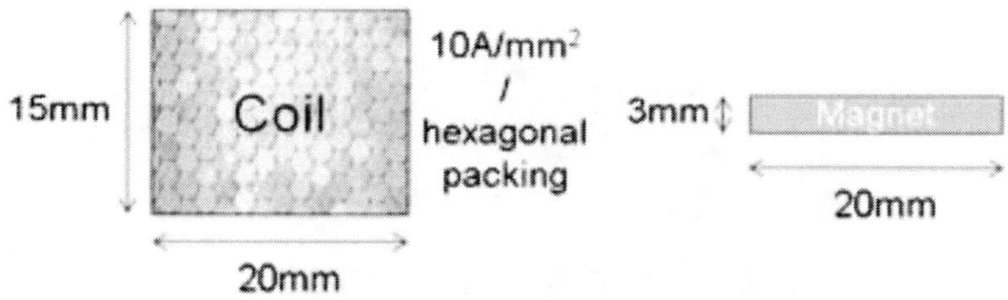

Figure 4. A simple comparison between the area of electromagnet coil and NdFeB magnet required to produce the same magnetic field.

Whatever the price of rare earth magnets, it is generally recognised that their elimination from electrical machines will lead to a reduction in costs. [8] reported that traction motors without NdFeB materials would have significantly lower materials costs (Fig. 6). However caution must be taken in this respect; it is possible that the use of other motor technologies may result in increased costs in other elements of the electric vehicle drivetrain (power electronics or batteries). If not carefully managed this could offset or even exceed any cost benefits linked to the removal of rare earth magnets from the electrical machine.

It has also been widely reported that the extraction and refinement of rare earth oxides is a potentially environmentally damaging process [9] and [10]. This aligns with research by the automotive industry (see Fig. 7) which suggests that NdFeB magnets may be, per unit mass, more damaging than other materials commonly used in electrical machines. If the breakdown of materials is estimated for a representative rare earth permanent magnet electric traction motor, it can be calculated that the NdFeB magnets may be responsible for perhaps 25% of the material related greenhouse gas emissions, despite being less than 5% of the motor by mass.

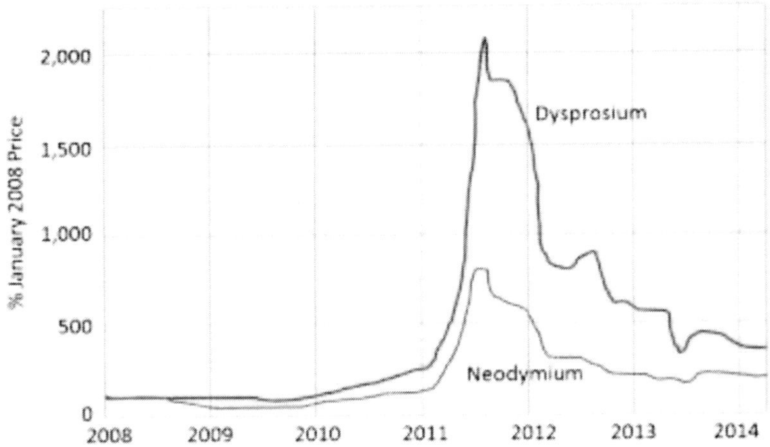

Figure 5. Price history for neodymium and dysprosium rare earth materials [7].

Whilst this analysis has the potential to exaggerate the environmental impact of NdFeB magnets when considered as part of the overall electric vehicle system, it is perhaps still true that their elimination could reduce the environmental footprint of traction motor manufacture, reducing that of electric vehicle manufacture in turn.

1.3. Other considerations with rare earth permanent magnet motors

Other important considerations relate to specific limitations in the performance of electrical machines which use these materials.

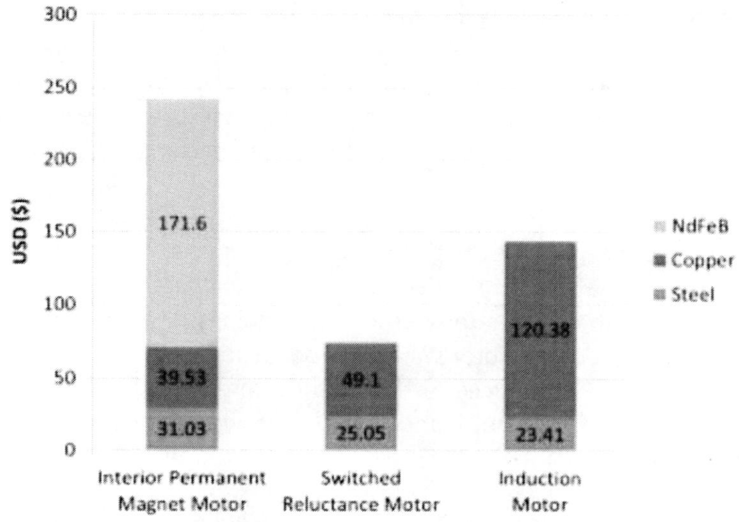

Figure 6. Materials cost in a 30 kW traction motor for a motor with rare earth magnets ('interior permanent magnet motor') and two options without [8].

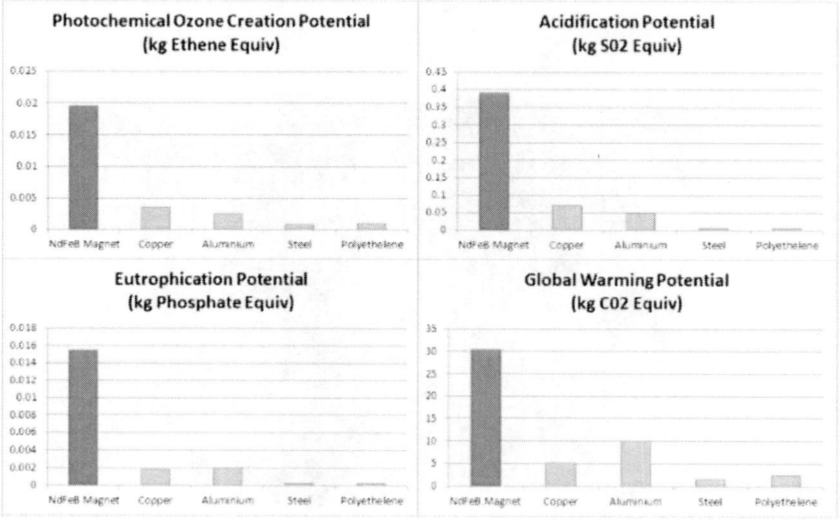

Figure 7. Showing lifecycle data for the production of the different materials used in electric motors (reference withheld at the request of the provider).

Rare earth magnets (along with other types of permanent magnet) are always ‘on’, whether the electrical machine where they are used is operational or not. This means that when a rare earth magnet (or other magnet based motor) is rotating, even with no electrical current applied, magnetic flux will still be present within the machine. This leads to a number of undesirable effects.

One consequence will be high, magnetically induced voltages at motor terminals if the motor's electrical terminals are open-circuit or, alternatively, high currents through motor windings if the motor terminals are short-circuited. The motor will also produce losses; these will be due to so-called iron losses, induced in the motor's core by the rotating magnet fields, and to resistive losses in motor windings, where the motor terminals are short-circuited.

Permanent magnet based motors will also ‘cog’ when not on load. Cogging is the term used to describe the way in which the motors, when not under load, may produce a fluctuating (though zero average) torque as they rotate. This cogging has been identified as a cause of noise, vibration and harshness problems in electric drive systems. This problem is often resolved by ‘skewing’ the motor rotor (Fig. 8), however skewing tends to result in reduced motor performance overall [11].

Figure 8. Skewed traction motor rotor, showing five steps in the rotor profile along its axial length.

Finally, a process known as 'field weakening' highlights a key disadvantage of magnet based motor technologies. At higher motor speeds, cost effective power electronic inverters must use field weakening in order to allow higher motor speeds to be achieved. As magnets spin in a stator they induce a voltage across motor windings known as the 'back' electromotive force (back EMF). With a permanent magnet, the scale of this back EMF voltage is directly proportional to motor speed. Therefore once the motor has reached a sufficient speed, the back EMF becomes so large that it becomes larger than the available motor supply voltage, preventing higher speed operation. This effect can be mitigated by field weakening the motor, literally injecting additional currents into the motor windings in a specific way in order to cancel magnet flux. This control strategy works well, perhaps extending motor speed by a factor of 3 to 5; however by requiring additional current to be injected into motor windings, it results in increased resistive losses and therefore lower efficiencies.

It is clear that, whilst rare earth magnetic materials enable electric motors with very high torque densities, their replacement offers attractions in terms of motor cost, environmental footprint and even some aspects performance.

2. ALTERNATIVES TO RARE EARTH MAGNETS IN ELECTRICAL MACHINES

It is therefore worthwhile to consider the alternatives to the use of rare earth permanent magnets, whether these aim to reduce the amount of rare earth material or substitute it completely.

2.1. Reduced rare earth

Whilst not allowing the complete elimination of rare earth magnet materials, much work has focussed on reducing the quantity of magnet used in traction applications.

Efforts in this area have tended to focus on two alternative approaches. Firstly magnet manufacturers are seeking to reduce the rare earth content of magnets whilst maintaining or increasing their performance. An example is Hitachi Metals [12] who have developed magnets which reduce

Dysprosium content when compared to conventional NdFeB materials, reportedly without a reduction in their high temperature coercivity. These magnets are manufactured using a new process, which involves the diffusion of Dysprosium into the magnet material in place of direct alloying. Others [13] are seeking to reduce the grain size in the magnets to nanoscale with the expectation that this will significantly increase the Maximum Energy Product of the material.

A second area has been to make use of recycled rare earth materials. Computer hard disk drives have for many years used rare earth magnets containing Neodymium and Dysprosium. These are reported to be available in their billions and techniques are being investigated to recover these materials; however there are concerns relating to the quality and repeatability of such recycled magnetic materials [14].

Finally there has been much work on designing electrical machines which are able to produce high torque densities whilst using less rare earth magnetic material. The most successful approaches in this regard have tended to focus on the use of so-called hybrid motor technologies. These make use of smaller amounts of magnetic materials embedded into 'salient' rotor structures (these are in many ways related to reluctance motors which are addressed later in this section). An example of this type of motor technology is used in the BMW i3 electric vehicle; this uses a highly salient rotor topology (Fig. 9) to produce 250 Nm torque and 125 kW power from just 1 kg of magnet (estimated by the authors). This compares well with earlier hybrid motor designs, such as that used in the Nissan Leaf, which achieves 280 Nm and 80 kW with an estimated 2 kg of magnetic materials [15].

The availability of software systems which allow the rapid optimisation of electrical machine geometries has been important in this respect, allowing magnetic material to be minimised without reducing motor performance. However care must be taken to ensure that magnets are protected from reverse magnetic fields which could lead to their permanent demagnetisation. For this reason designs must ensure that short circuit currents induced during fault conditions do not become unacceptably high.

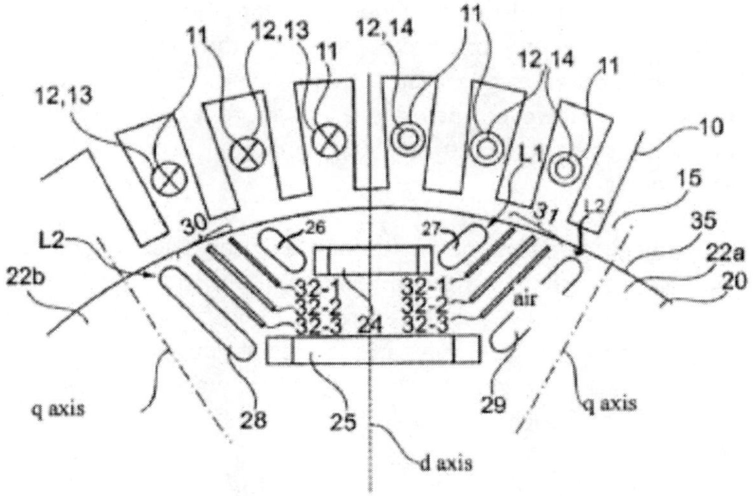

Figure 9. Showing the BMW hybrid motor technology which has allowed the volume of rare earth magnetic material to be significantly reduced [16].

2.2. Other magnetic materials

There has been much work reported in the development of magnetic materials not using Neodymium and Dysprosium. There have been reports of new, very high performance magnetic materials such as Iron Nitride [17], Manganese based compounds [18] and others [19]. However it is not clear how practical or close to market these materials are and it can be argued that commercialisation is likely to take several years, even once significant improvements have been achieved.

A number of magnetic materials developed prior to NdFeB have also been reconsidered; the relative performance of these materials is lower than NdFeB, as shown inFig. 2 and Fig. 3 earlier in the paper, however lower cost and better security of supply may be factors in their favour.

Aluminium Nickel Cobalt (AlNiCo) was developed in the 1930s, however whilst these magnets offer high remnant flux densities, their very low coercivity, and therefore the high risk of demagnetisation, makes them unattractive for use in high power density electrical machines. Researchers at Ames University in the US are focussing on coercivity improvements to these materials [20].

Samarium Cobalt (SmCo) magnets were developed in the 1970s and have frequently been used in aerospace applications; they are attractive as they can withstand higher temperatures than NdFeB [21] without suffering demagnetisation. However they are as or more expensive than NdFeB magnets and also contain the rare earth material Samarium, along with expensive Cobalt.

The final choice is perhaps the ferrite magnet, used since the 1950s. These are manufactured from Iron Oxide combined with the metals Strontium, Barium or Cobalt. In order to increase their remnant flux density, companies such as TDK [22] have had success with introducing quantities of Lanthanum; this is also a rare earth metal, however Lanthanum based ferrite costs remain much lower than for NdFeB.

The major challenges of a ferrite based motor design concern the low remnant flux density and low coercivity of these magnets. Having only one third of the remnant flux density of NdFeB magnets, to obtain a competitive torque density requires substantial flux focusing (concentrating the flux from multiple magnets), coupled with a design which also offers reluctance torque (see later section on reluctance motors for details). Equally, the coercivity of ferrite magnets is about one fifth to one third of the NdFeB magnets; as a result very careful design of the motor is necessary to withstand demagnetizing fields during field weakening and under short circuit conditions. One advantage, however, is that the magnets have a characteristic that their coercivity increases with temperature, being the opposite behaviours observed in materials such as NdFeB, and making them less sensitive to sudden demagnetisation in demanding applications. There has been much recent research into the use of ferrite magnets in electrical machines. An example is shown in Fig. 10[23], where a consortium of European organisations has constructed a 60 kW traction machine using ferrite magnets in a spoke type configuration. Other projects developing similar motors are also known to be ongoing in other countries at time of writing, with GE also designing a similar machine [24] and the authors also having successfully demonstrated this technology. It is yet to be seen how successful these motor designs will be, however initial findings are that they may offer a credible, low cost alternative.

Figure 10. A 60 kW ferrite magnet based traction motor.

2.3. Induction machines

Although induction machines are very common in industrial automation, they are less frequently used in automotive traction applications. The key exception is the Tesla Motor Corporation who has reported the use of copper rotor cage induction machines in all of their electric vehicles. Tesla also licenced this technology to Toyota, who used it in the now discontinued RAV4 EV SUV [25], and subsequently Tesla have reportedly made all of their related patents licensable at zero cost [26].

Induction motors contain no permanent magnetic materials, instead they operate by inducing electrical currents in conductors in the motor's rotor; these currents in turn give rise to a magnetic field in the rotor and thus produce torque. As a result, when switched off, these motors are inert, producing no electrical voltage or current, no losses and no cogging torque. These machines can produce high levels of performance using modern and well understood vector control techniques.

However, induction machines incur losses in their rotor conductors, which can result in total rotor losses typically two to three times higher than in a permanent magnet based motor. High rotor losses are not desirable as the rotating rotor is much more difficult to cool than the stationary stator. In practice these high losses mean not just that this type of electrical machine may be less efficient than other options [8], but also that in operation it may quickly become overheated. As a result it may be necessary to implement control strategies which temporarily enforce power restrictions in order to allow motor temperatures to recover to acceptable levels.

Improvements can be achieved through the use of a copper rotor cage (as is the practice for Tesla, for example Fig. 11), as opposed to the more commonly used aluminium. However the cost of manufacture of these copper based rotors will be higher; this is because of the high cost of copper, when compared to aluminium, and due to its higher melting temperature, which makes process control more demanding than for aluminium. Equally copper is a dense material, increasing motor mass. However this motor technology has undoubtedly been a success for Tesla in recent years.

Figure 11. Induction motor rotors with copper rotor cage.

2.4. Wound rotor motors

Electrical machines with wound rotors are the basis for almost all electrical power generation worldwide where, at multi-megawatt power levels, they offer extremely high efficiency and reliability. In motor applications, an AC rotating field is applied to the stator motor windings (in the same way as for both induction and permanent magnet based motors). In place of permanent magnets on the rotor, copper windings are used to set-up a magnetic field, with an electrical current being supplied to these rotating windings from the stationary part of the machine. This requires a mechanism for passing current from the stationary part of the motor to rotating part of the rotor. This is often achieved through the use of slip rings; these use stationary carbon brushes running on a rotating conductive ring, mounted in the motor to provide a conductive path to the rotor. However slip rings can prove unreliable with brushes wearing over time. Brush wear can lead to a build-up of carbon dust in the motor which can lead to electrical faults. This limitation can potentially be overcome by the use of wireless power transfer based systems, for example [27] and [28], which are similar to the wireless charging systems which are now often used to charge consumer electronic devices. However these systems are currently immature and may by their nature be less efficient.

A further drawback is that additional switching devices are needed in the power electronic inverter in order to provide controlled current to the rotor; this increases system complexity and therefore cost. Also, in a similar way to induction machines, heat extraction from rotor conductors may become a limiting factor in practice, with again careful control needed to prevent overheating.

Car manufacturer Renault, working with German company Continental (Fig. 12) [29], have successfully used this type of motor in cars including the Renault Fluence and Renault Zoe.

2.5. Reluctance motors

There is also the option to use a family of motors which make use of the so-called reluctance principle to generate torque. The generation of reluctance torque is based on the fundamental tendency for electromagnetic flux to seek the path of least resistance; this is observed by the way in which a permanent magnet is attracted to a ferrous material, such as steel. Here the magnetic flux generated in the permanent magnet seeks out the steel as this provides a ‘low reluctance’ path along which it can flow. This effect can also be used in an electrical machine, as one example using electromagnets in the stator to consecutively attract ferrous teeth in the machine's rotor (Fig. 13).

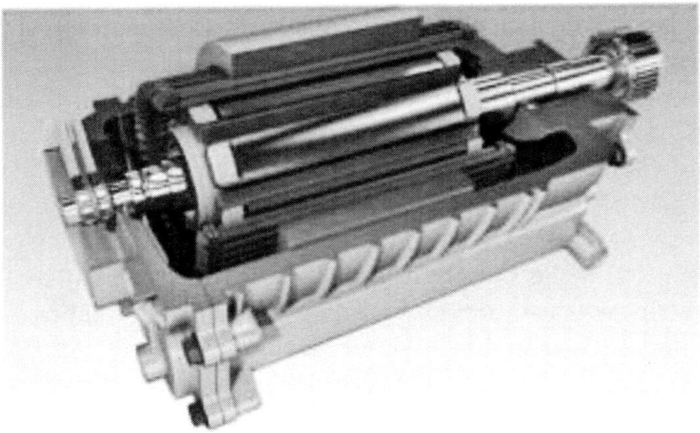

Figure 12. Continental wound rotor synchronous motor.

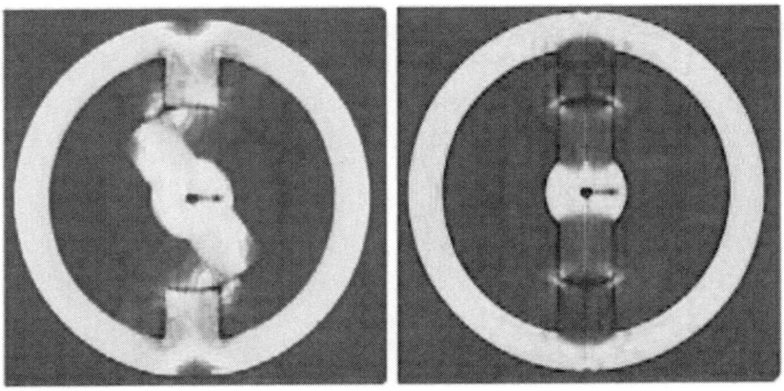

Figure 13. Finite element plots showing the reluctance motor principle of operation: (left) when the rotor (inner portion) is not aligned with the stator (outer portion) a magnetic force results which brings the two into alignment (right).

Two types of reluctance motor are possible. The first, the synchronous reluctance motor, conventionally uses a stator very similar to that used in an induction motor. This stator is excited by an inverter drive applying sinusoidal currents to one motor phase after another, generally forming a three phase system. The motor's rotor is salient, offering a low reluctance in one axis and a high reluctance in the other axis, therefore allowing the motor to lock synchronously onto the rotating stator field. These types of machines have the potential for very high efficiency, with ABB having launched a range of very efficient industrial machines using this technology in 2011 [30]. These machines have a number of attractions; however they suffer from relatively low torque densities which may be a limiting factor in electric vehicle applications.

In contrast the switched reluctance motor produces torque using a subtlety different mechanism. The rotor still has saliency, interspersed regions of high and low reluctance. However in this case control is provided by a less common form of power converter, known as an asymmetric half bridge converter. The general unfamiliarity and the lack of commercial availability of these converters has been a limiting factor in the uptake of these machines. However they have been shown to be very inexpensive [8] and robust, producing competitive torque densities [31]. The authors have constructed motors of this type which have demonstrated performance very close to that of existing rare earth magnet based motors (Fig. 14).

Figure 14. 75 kW switched reluctance motor for use in hybrid electric truck application, acting as a directly replacement for a NdFeB based motor drive.

Concerns have been raised with switched reluctance motors relating to their tendency to produce high torque ripple and acoustic noise when compared with other motor types, undesirable in modern and refined vehicles, however it is likely these risks can be mitigated through careful motor, control and system design.

3. TECHNOLOGY COMPARISON

This analysis raises the question of which traction motor technology is the best. However this is a question which probably misses the point; as with the availability of a variety of internal combustion engine technologies today (gasoline vs diesel, supercharged vs normally aspirated) there is almost certainly room for different technologies. Table 1summarises the differences between technologies, based on the sources already cited in the proceeding sections and the authors' own research.

It is perhaps ferrite magnet and switched reluctance motors which may offer the lowest cost in volume manufacture, though care must be taken not to increase system costs (power electronic converter and battery) and neither technology is yet fully proven in this application. However rare earth permanent magnet motors are likely to continue to have a place in very high performance applications, for example where individual electric motors are placed in each vehicle wheel, and torque density is therefore the critical requirement. Each of these technologies today has its proponents and detractors, however it is also likely true that any of these technologies can be made to work; it is simply necessary to apply the time and money to achieve a level of design maturity commensurate with use in a volume produced electric vehicle.

4. CONCLUSIONS

The rare earth magnetic material, Neodymium Iron Boron, forms the basis for the traction motors used in many of todays leading Battery and Hybrid Electric Vehicles. These magnets enable the design of motors which offer extremely high torque densities, making them compact and lightweight, whilst also offering high efficiencies.

Table 1. Comparison of electric motor technologies which reduce or eliminate rare earth magnets.

Motor technology	Reduced NdFeB magnet	Ferrite permanent magnet	Copper rotor induction	Wound rotor synchronous	Switched reluctance
Peak power	80 kW	80 kW	50 kW	50 kW	75 kW
Peak efficiency	98%	96%	96%	96%	97%
Active material cost	$223	$154	$144	$144	$118
Active material cost per kW	$2.78/kW	$1.93/kW	$2.88/kW	£2.88/kW	£1.57/kW
Torque density	15 Nm/kg	11 Nm/kg	10 Nm/kg	10 Nm/kg	15 Nm/kg

However there are a number of arguments that this technology may not offer the best long-term solution for use in this application. In particular rare earth magnets are expensive, doubling or more the raw material cost of the electric motor, whilst perhaps not being particularly sustainable in terms of their mining and refinement. NdFeB motors may also not be as efficient during normal vehicle operating conditions as the headline claims may indicate, with control strategies needed to weaken the influence of the magnets, allowing higher speed operation, but being a source of inefficiency.

Even if these factors can be somewhat moderated, there is nevertheless a strong argument to investigate the exploitation of other motor technologies which do not need rare earth magnetic materials. Renault and Tesla have already employed wound rotor and induction motor technologies respectively, eliminating rare earth magnets. These and other technologies, notably switched reluctance motors and those which replace rare earth magnets with low cost ferrites, can perhaps form the basis of even higher performance traction motors in the future.

REFERENCES

1. Y. Matsuura, M. Sagawa, et al., "Process for producing permanent magnet materials," US Patent 4,597,938, 1986.

2. H.R. Kirchmayr, Permanent magnets and hard magnetic materials, J. Phys. D. Appl. Phys. 29 (1996) 2763–2778.

3. (2015, 26/02/2014). Arnold Magnetics, ProductsAvailable http://www.arnoldmagnetics.com/Products.aspx.

4. (2011, 26/02/2015). TDK., Ferrite magnetsAvailable http://product.tdk.com/en/catalog/datasheets/e321.pdf.

5. P. Tenaud, H. Lemaire, et al., Recent improvements in NdFeB sintered magnets, J. Magn. Magn. Mater. 101 (1991) 328–332.

6. K. Bourzac, The rare-earth crisis, MIT Technology Review2011. (Available: http:// www.technologyreview. com/featuredstory /423730/the-rare-earth-crisis/).

7. J. Rowlatt, Rare earths: neither rare, nor earthsAvailable: http://www.bbc.co.uk/ news/magazine-26687605 2014.

8. D.G. Dorrell, A.M. Knight, et al., Comparison of different motor design drives for hybrid electric vehicles, Energy Conversion Congress and Exposition (ECCE), 2010 IEEE, 2010, pp. 3352–3359.

9. Anon., "Rare-earth mining in China comes at a heavy cost for local villages," in Guardian, ed. London, 2012.

10. S. Parry, E. Sand Douglas, "In China, the true cost of Britain's clean, green wind power experiment: pollution on a disastrous scale," in Daily Mail, ed. London, 2011.

11. K.J. Binns, F.B. Chaaban, et al., Major design parameters of a solid canned permanent magnet motor with skewed magnets, Electr. Power Appl. IEE Proc. B 140 (1993) 161–165.

12. R. Gehm, Hitachi Metals Reduces Rare-earth Dysprosium in Electric-motor Magnets, SAE International, 2013. (Available: http://articles.sae.org/11988/).

13. S. KOZAWA, Trends and Problems in Research of Permanent Magnets for Motors, Science and Technology Trends2010. (Available: http://data.nistep.go.jp/dspace/ bitstream/11035/2854/1/NISTEP-STT038E-40.pdf).

14. K. Binnemans, P.T. Jones, et al., Recycling of rare earths: a critical review, J. Clean. Prod. 51 (2013) 1–22.

15. Y. Sato, S. Ishikawa, et al., Development of high response motor and inverter system for the Nissan LEAF electric vehiclePresented at the SAE 2011 World Congress & Exhibition, Detroit, Michigan, United States, 2011.

16. J. Merwerth, J. Halbedel, et al., "Electrical Drive Motor for a Vehicle," ed: Google Patents, 2012.

17. J.P. Wang, S. He et al., "Iron nitride permanent magnet and technique for forming iron nitride permanent magnet," ed: Google Patents, 2014.

18. J.M.D. Coey, New permanent magnets; manganese compounds, J. Phys. Condens. Matter 26 (2014) 064211.

19. R. Filip, B. Sam, Rare earth replacement magnets, J. Phys. Condens. Matter 26 (2014) 060301.

20. L. Zhou, M.K. Miller, et al., Architecture and magnetism of alnico, Acta Mater. 74 (2014) 224–233.

21. F.F. Westendorp, On the coercivity of SmCo5, Solid State Commun. 8 (1970) 139–141.

22. TDK, TDK High Performance Ferrite Magnets, FB series, 2014. (Available: http:// product.tdk.com/en/ techjournal/ archives/ vol07_ fb/contents03.html).

23. MotorBrain projectAvailable http://www.motorbrain.eu/.

24. S.J. Galioto, P.B. Reddy, et al., Effect of magnet types on performance of high speed spoke interior permanent magnet machines designed for traction applications, Proc. IEEE ECCE, 2014, pp. 4513–4522.

25. J. Voelcker, Tesla deal to supply Toyota RAV4 EV powertrain to end this yearAvailable: http://www.greencarreports. com/news/ 1092023_ tesla-deal-to-supply-toyota-rav4-ev-powertrain-to-end-this-year 2014.

26. E. Musk, All Our Patent Are Belong to YouAvailable: http:// www.teslamotors.com/ blog/all-our-patent-are-belong-you 2014.

27. J. Veitengruber, F. Rinderknecht, et al., Preliminary investigations of an inductive power transfer system for the rotor power supply of an electric traction drive, Ecological Vehicles and Renewable Energies (EVER), 2014 Ninth International Conference on, 2014, pp. 1–8.

28. D.C. Ludois, J.K. Reed, et al., Capacitive power transfer for rotor field current in synchronous machines, Power Electron IEEE Trans. 27 (2012) 4638–4645.

29. Continental Provides a Broad Range of Electric Drives for Electric Vehicles, 2010. (Available: https://www.conti-online.com/www/ pressportal_com_en/themes/ press_releases/3_automotive_group/ powertrain/press_releases/pr_2010_04_29_ elektromotoren_en.html).

30. A. Tammi, IE4 synchronous reluctance motor-drive package on sale this weekAvailable: http://www.abb-conversations.com/2012/11/ie4-synchronous-reluctance-motor-drive-package-on-sale-this-week/ 2012.

31. Y. Takano, M. Takeno, et al., Design and analysis of a switched reluctance motor for next generation hybrid vehicle without PM materials, Power Electronics Conference (IPEC), 2010 International, 2010, pp. 1801–1806.

Chapter 7

The Application of Electric Drive Technologies in City Buses

Zlatomir Živanović[1] and Zoran Nikolić[2]

[1] University of Belgrade, Institute of Nuclear Sciences VINCA, Belgrade, Serbia

[2] Institute of Technical Sciences of the Serbian Academy of Sciences and Arts, Belgrade, Serbia

1. INTRODUCTION

The environmental concerns and limited reserves of fossil fuels have generated an increased interest for alternative propulsive systems of vehicles. On the other hand, vehicle manufacturers are increasingly facing demands for reducing emissions of harmful gases by the vehicles in accordance with the increasingly stringent legislation.

Buses as means of public transportation could reduce considerably the problems caused by the traffic in the urban areas through the usage, among other things, innovative techniques and technologies of vehicle propulsion systems.

The development of innovative technologies is increasingly oriented towards electrification of vehicle propulsion systems expected to lead to: a reduction of harmful emissions, an increased efficiency of vehicles, improved performances, a reduction of fuel consumption, a reduction of noise, and potentially lower maintenance costs. An electric drive technology implies a technology employing at least one drive device called electric motor. Three key electric drive technologies are: hybrid electric, battery electric, and fuel cell electric technologies.

In this chapter the application of electric technologies in the bus propulsion systems is considered through: an analysis of the state of development of city buses, an analysis of the advantages and shortcomings of electric drive technologies, and identification of the problems standing in the way of their greater commercialization. The presented examples of the developed city buses describe the basic characteristics of the applied propulsion systems and their advantages. The development of hybrid technologies for bus propulsion has grown considerably over the past several years. These technologies have reached massive applications in North America and their expansion to Europe has been initiated during the past several years.

In the part of the chapter dealing with hybrid electric buses a typical bus driving system and its components are reviewed and hybrid systems of the major world manufacturers are presented. Special attention is paid to the comparative analysis the hybrid electric buses and conventional and CNG buses. The available literature data have been critically processed and re-presented. Finally, the characteristics of hybrid city buses of some American and European manufacturers which have found the most widespread applications are reviewed.

The fuel cell powered buses draw special attention of users owing the efficiency of their propulsion system and their ability to cut drastically harmful emissions. Even though they are still not widely used, judging by the number of demonstrated projects the development of fuel cell buses is very intensive throughout the world. Barriers to their wider use are very high costs, lack of an adequate infrastructure, and relatively small radius of movement.

Fuel cell buses are vehicles with zero emissions. CO_2 emissions depend on the type and method of production of fuel for fuel cells. In the part of the chapter dealing with fuel cell buses the typical bus configuration, its subsystems, its ecological characteristics, and costs are reviewed. Some of the development projects and characteristics of a new generation of fuel cell buses are presented.

Battery electric technologies are among technologies which reduce drastically the impact of a vehicle on the environment, however, they are still far from the proven technologies. The reason is the current level of developing technology of the energy storage devices for these vehicles. Influence of the batteries on commercialization of these buses is more pronounced compared to the other electric drive buses. A significant advancement in the area of energy sources has been made over the past several years by the development of lithium-ion batteries which lead to the development of an increased number of prototypes, even to a series production of these vehicles. In the part of the chapter dealing with battery electric buses the characteristics of some of these realized buses are reviewed.

In a separate part of this chapter the characteristics of energy storage device for the electric propulsion systems of the realized buses are presented, and the expectations from further development trends of the energy source devices are outlined.

2. AN OVERVIEW OF ELECTRIC DRIVE TECHNOLOGIES

Depending on the degree of electrification of propulsion system, Figure 1, three key electric drive technologies for power the electric vehicles are: hybrid electric, battery electric and fuel cell electric technologies [1].

Hybrid electric technology: A hybrid electric technology uses both an electric motor (EM) and an internal combustion engine (ICE) to propel the vehicle. Vehicles equipped with this technology are called Hybrid Electrics Vehicles (HEVs).

As can be seen from Figure 1, source of energy to power the vehicle with hybrid electric technologies are fuels, including alternative, which can be used in IC engines and electricity stored in the batteries or ultra capacitors. The charge energy storage to electricity is performed via IC engine and/or via the regenerative braking.

Special case of hybrid electric technology is plug-in hybrid electric (PHE) technology. It has a battery that can be charged off board by plugging into the grid and which enables it to travel certain kilometres solely on electricity. Vehicles equipped with this technology are called Plug-in Hybrid Electrics Vehicles (PHEVs).

Battery electric technology: Battery electric technology uses a relatively large on-board battery to propel the vehicle. Battery provides energy for propulsion through an electric traction motor(s) as well as power for all vehicle accessory systems. Some electric vehicles (EVs) can use to drive auxiliary devices like an on board generator, which makes them have characteristics of hybrid solutions. Vehicles equipped with this technology are called Battery Electrics Vehicles (BEVs).

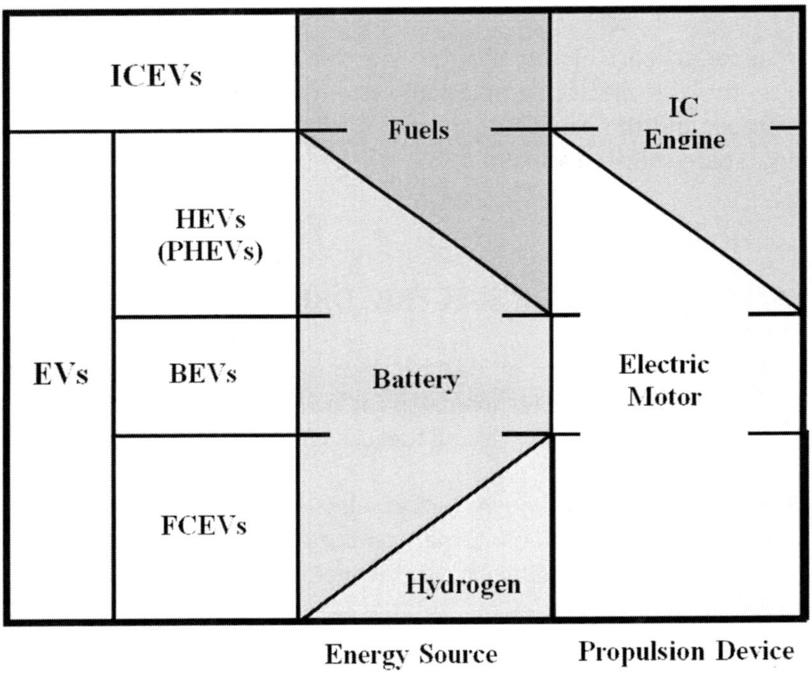

Figure 1. Different degrees of electrification of road vehicles.

Fuel cell electric technology: Fuel Cells are energy conversion devices set to replace combustion engines and compliment batteries in a number of applications. They convert the chemical energy contained in fuels, into electrical energy (electricity), with heat and water generated as by-products. Fuel cells continue to generate electricity for as long as a fuel is supplied, similar to traditional engines. However unlike engines, where fuels are burnt to convert chemical energy into kinetic energy, fuel cells convert fuels directly into electricity via an electrochemical process that does not require combustion. Vehicles equipped with this technology are called Fuel Cell Electrics Vehicles (FCEVs).

Electric drive technologies also, usually, incorporate other technologies, which reduce energy consumption, for example regenerative braking. That allows the electric motor to re-capture the energy expended during braking that would normally be lost. This improves energy efficiency and reduces wear on the brakes.

2.1. Electric Drive Technology Configurations

Hybrid electric drive configurations consist of a fuel-burning prime power source – generally an ICE–coupled with an electrochemical or electrostatic energy storage device. These two power sources work in conjunction to provide energy for propulsion through an electric traction drive system. Power for all vehicle accessory systems can be provided electrically or mechanically from the ICE or combinations of both. There are currently many different hybrid-electric system designs utilizing ICE, alternative fuels engines, gas turbines or fuel cells in conjunction with batteries. These design options are grouped in three categories: series, parallel and series-parallel configurations.

A series hybrid-electric drive system, Figure 2, consists of an engine directly connected to an electric generator (or alternator). The arrows indicate the mechanical and electrical energy flow.

Power from the generator is sent to the drive motor and/or energy storage batteries according to their needs. There is no mechanical coupling between the engine and drive wheels, so the engine can run at a constant and efficient rate, even as the vehicle changes speed. The serial hybrid technology is the most common hybrid technology.

In a parallel hybrid-electric drive system, Figure 3, both of the power sources (engine and electric motor) are coupled mechanically to the vehicle's wheels. In different configurations, the motor may be coupled to the wheels either through the transmission (pre-transmission parallel design) or directly to the wheels after the transmission (post-transmission parallel design). Each of these configurations has its advantages.

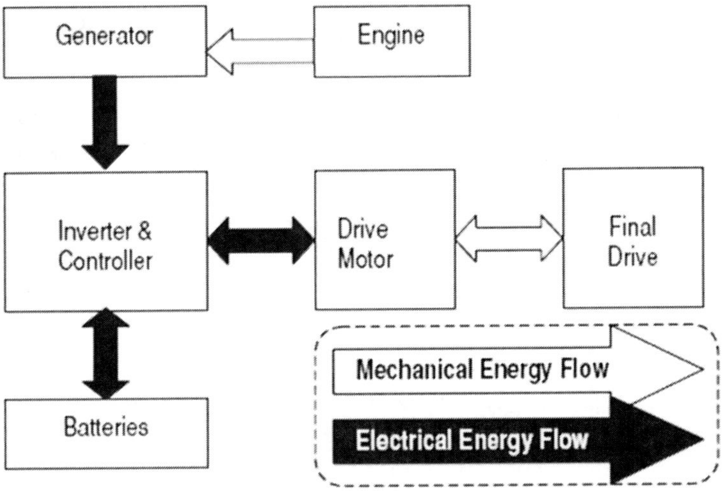

Figure 2. Series hybrid electric drive system.

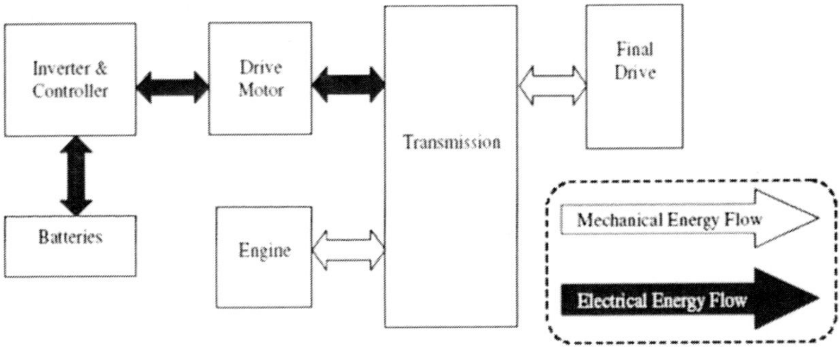

Figure 3. Parallel hybrid electric drive system.

A series-parallel design, also known as power-split or dual mode hybrid system, is interesting because with proper control strategy it can be designed to take advantage of both parallel and series types and avoid their drawbacks.

Battery electric drive system are illustrated in Figure 4. The two types of system configurations are possible depending on the positioning and size of the electric motors.

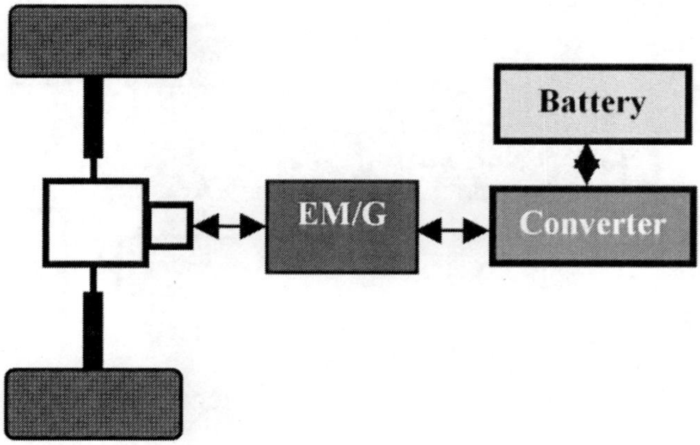

Figure 4. Battery electric drive system.

The central motor type is currently more common. However, the requirement to transfer power from the motor to the wheels does involve some losses in efficiency through friction.

The hub motor type can avoid many of the transmission losses experienced in the central motor type, but are a less regularly used technology.

In fuel cell drive system fuel cell is providing the electric energy needed to run of vehicles. There are two types of fuel cell drive configuration:

- Fuel cell drive without energy storage device (non-hybrid fuel cell vehicles) and

- Fuel cell drive with energy storage device (hybrid fuel cell vehicles), Figure 5.

Fuel cell hybrids operate much like other hybrid electric vehicles but with fuel cells producing electricity that charges the batteries, and a motor that converts electricity from the batteries into mechanical energy that drives the wheels.

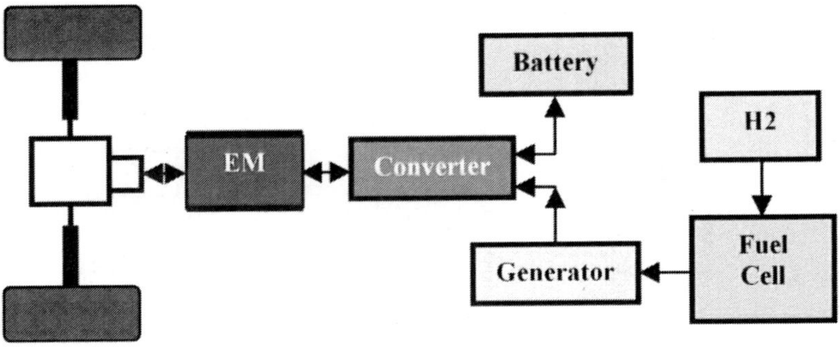

Figure 5. Fuel cell drive system with energy storage device.

2.2. Advantages And Disadvantages Of Electric Drive Technologies

There are some major advantages of electric drive technologies but there are also some disadvantages.Table 1 summarizes the advantages and disadvantages of a hybrid-electric, plug-in hybrid electric battery and fuel cell drive systems [2]:

3. HYBRID ELECTRIC BUSES

At the IAA 1969 in Frankfurt, Daimler presented the first electric test bus– an early example of hybrid drive technology. In 1979, Daimler has launched a five-year model trial with a total of 13 Mercedes-Benz OE 305 electric-diesel hybrid buses in regular service. Since 2008 Orion hybrid buses are in regular service on the roads of major U.S. cities, but from 2009 Mercedes-Benz Citaro BlueTec Hybrid is in daily operation [3].

The first sales of serial hybrid city buses in Japan began 1991, when Hino delivered test buses in eight cities.

From 1997 until today leaders by the number of hybrid buses in commercial use are the United States and Canada.

Table 1. Advantages and disadvantages of electric drive technologies.

Technology	Advantages	Disadvantages
Hybrid electric	Lower fuelling costs; Reduced fuel consumption and tailpipe emissions; Recovered energy from regenerative braking	Higher initial cost; Complexity of two power trains; Component availability
Plug-in hybrid electric	Cleaner electric energy thanks advanced technologies or renewable; Reduced fuel consumption and tailpipe emissions; Optimized fuel efficiency and performance; Recovered energy from regenerative braking; Grid connection potential; Pure zero-emission capability	Higher initial cost; Complexity of two power trains; Component availability-batteries, powertrains, power electronics; Cost of batteries and battery replacement; Added weight
Battery electric	Use of cleaner electric energy; Zero tailpipe emissions; Overnight battery recharging; Recycled energy from regenerative braking; Lower fuel and operational costs; Quiet operation	Mileage range; Battery technology still to be improved; Possible need for public recharging infrastructure

Fuel cell	Zero tailpipe emissions; Higher energy efficiency than the IC engine; Recovered energy from regenerative braking; Potential of near-zero well-to-wheel emissions when using renewable fuels to produce hydrogen; No dependence on petroleum	Higher initial cost; Increased reliability and durability; Hydrogen generation and onboard storage; Availability and affordability of hydrogen refueling; Codes and standards development; Scalability for mass manufacture;

3.1. Hybrid Electric Bus Architectures

A hybrid electric bus (HEB) usually combines an internal combustion engine with the battery and an electric motor. The ICE can be fueled by gasoline, diesel, or other (natural gas, biofuel) and work either in series or in parallel with the electric motor. Regenerative braking capability in HEBs minimizes energy losses by recovering some of the kinetic energy used to slow down or stop a vehicle.

In a series hybrid configuration [4], Figure 6, the ICE drives a generator to feed the electric motor and recharge the battery. Braking energy can be captured and stored in the battery ("regenerative braking"). The engine can be downsized compared to a conventional drive-train with the same performance, meaning lower ICE weight and higher energy efficiency.

The electric motor powers the drive system, using either energy stored in batteries, or from the engine, or from both as needed. The engine is more efficient at lower speeds and higher load, so the series hybrid is preferred for slow and start-and-stop city driving.

In a parallel hybrid configuration both the engine and the electric motor are linked to the transmission so that either of them, or both at the same time, may provide the power to turn the wheels. Since the parallel hybrid configuration allows the engine to drive the wheels also through a direct mechanical path, it offers better efficiency than a series hybrid configuration, and a more functional and flexible design.

3.2. Hybrid Electric Bus Components

The principal hybrid-electric bus components include:

1) an Auxiliary Power Unit (APU),

2) a drive motor,

3) a controller and inverter,

4) an energy storage device and

5) other auxiliary systems, such as air conditioning and lighting.

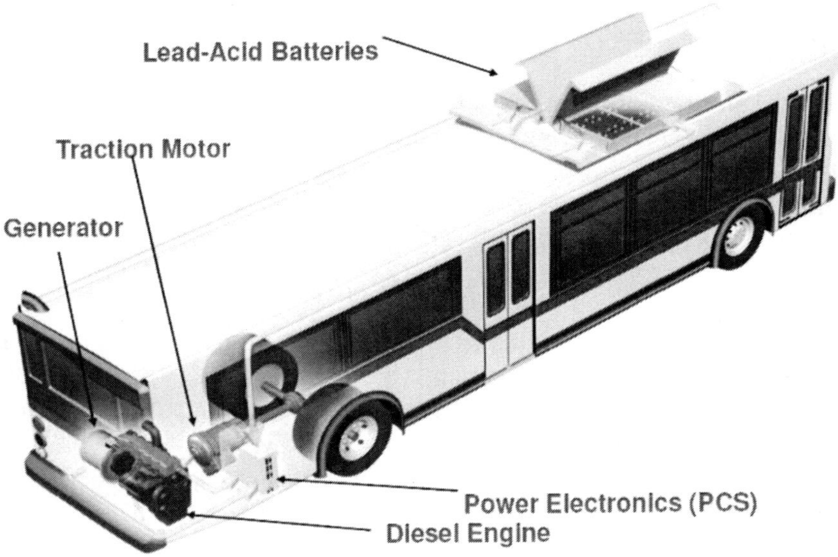

Figure 6. Series hybrid propulsion system.

Auxiliary Power Units: Auxiliary Power Units (APUs) used in hybrid-electric buses are available in a number of configurations including IC engines, fuel cells, and with different fuels, such as diesel, gasoline and compressed natural gas (CNG), liquid natural gas (LNG) and propane. The engine is typically sized for the average bus power demand, not peak power demand since the energy storage device provides supplementary power. Engines in hybrid configurations also operate over a narrower range of load and speed combinations compared to engines in conventional buses.

Drive motors: Two primary types of electric motors can be used in electric vehicles, direct current (DC) motors and alternating current (AC) motors. On a power comparative basis, an AC motor generally exhibits higher efficiency, has a favourable power to size/weight ratio, is less expensive and generates regenerative braking energy more efficiently than a DC motor. Electric drive motors are connected to the vehicle wheels either directly, referred to as wheel motors, or through a transmission and differential assembly. Wheel motors are more efficient both in drive cycle and in the regenerative cycle by eliminating the losses in the mechanical transmission and the differential. However, wheel motors are expensive.

Controller and inverter: The electronic controller regulates the amount of energy, (DC power in the case of batteries), that is transferred or converted to AC power by the inverter (in AC motors) for acceleration. It also ensures that voltage is maintained within the specifications required for operating the motor. An electronic controller can also recover electrical energy by switching the motor to a generator in order to capture the vehicle's kinetic energy via regenerative braking. The controller also ensures that the regenerative current does not overcharge a battery.

Energy storage devices: Energy storage devices provide necessary of the energy in hybrid-electric buses to supplement the APU energy when there is a high demand (e.g., acceleration from stop, speed acceleration, climbing an up-hill gradient) and to recover and store the energy generated during deceleration (e.g., braking, down-hill coasting).

3.3. Major Manufacturers Of Hybrid Systems

The major manufacturers of hybrid systems are shown in Table 2:

BAE Systems is a major integrator and supplier of integrated hybrid electric propulsion for Orion hybrid electric buses manufactured by Daimler Buses North America. BAE Systems produces the HybriDrive series propulsion system. The HybriDrive is composed of a traction motor and a traction generator to provide power to the vehicle. The liquid cooled, high power-to-weight ratio electric traction motor connects directly to a standard drive shaft and rear axle to provide traction power and regenerative braking. HybriDrive series system powers more than 3.500 buses [5].

Allison has developed EP50 parallel hybrid electric propulsion system with two motors capable of producing 75 kW of continuous power and up to 150 kW of power at full potential. Although designed as a parallel architecture, a power-split or two-mode hybrid electric bus can operate in either a series, or a parallel configuration. There are more than 4,600 buses with the Allison Two Mode Parallel Hybrid Systems in operation across 216 cities and 9 countries. The buses have driven more than 600 million kilometres, saved over 75 million litres of fuel and eliminated more than 197.000 metric tones of CO_2 [6].

Table 2. Major manufacturers of hybrid systems.

Manufacturer	Propulsion system	Type	Country
BAE	HybriDrive	Series	USA
Allison	E^{P}40/E^{P}50	Series-parallel	USA
ISE	ThunderVolt	Series	USA
Siemens	ELFA	Series	Germany
Eaton	EHPS	Parallel	USA
Volvo	I-SAM	Parallel	Canada
Voith	DIWAhybrid	Parallel	Germany

ISE Corporation produces the ThunderVolt series drive system who has five key subsystems: Motive Drive Subsystem (electric drive motors, motor controller, gear reduction system, and related components); Auxiliary Power Unit Subsystem (engine, electric generator, and related components); Energy Storage Subsystem (integrated pack of either batteries or ultracapacitors); Vehicle Control and Diagnostics and Electrically-Driven Accessories (electrical power steering and braking systems, air conditioning systems, and related components). ISE components for electric and hybrid drive are manufactured by Siemens (ELFA). ThunderVolt packs have been integrated in over 300 in-service buses, and operated for over 10 million miles (16,09 million km) cumulative [7].

Siemens ELFA hybrid propulsion system was initially developed in the mid 90's for diesel electric buses. To date, Siemens has outfitted more than 700 hybrid busses in the U.S., Italy, Germany and Japan. More than 6 Million operation hours attest to the reliability of the system [8] Rugged liquid-cooled induction motors with power ratings from 50 kW to 180 kW with reduction gearboxes are used as standard for ELFA traction systems. Permanent-magnet generators are used for all of the latest ELFA traction drive generation. The traction converters play a key role in ELFA traction systems. The complete ELFA traction system is controlled using just one standard traction converter software.

Eaton hybrid power system (EHPS) uses a parallel configuration [9]. EHPS consists of an automated clutch, electric motor/generator, motor controller/inverter, energy storage unit, automated manual transmission and an integrated supervisory hybrid control module, takes energy created during braking and regenerates it for later use. An electric motor/generator is located between the output of an automatic clutch and the input to an automated mechanical transmission. The electric motor's peak output is 44 kW.

Volvo has developed a parallel hybrid system [10] with integrated starter alternator motor (ISAM) system that can be used across the complete product range. ISAM is located between the diesel engine and the gearbox and facilitates a compact parallel hybrid package. ISAM integrates the starter motor, electric motor, generator and electronic control unit into one component. The electric motor is used to start and accelerate the bus up to about 20 km/h, while the diesel engine takes over at higher speeds.

Voith presented its parallel DIWAhybrid system at the American Public Transportation Association (APTA) show in 2008. DIWAhybrid system builds on the proven DIWA automatic transmission and is designed for up to 290 kW power input and a maximum input torque of 1.600 Nm. At 150 kW electric traction power, the DIWAhybrid system reduces the load of the diesel engine enough for the latter to be substantially smaller than on conventional diesel buses [11]. Voith has begun production of DIWAhybrid drive system for city buses in the USA at the end of 2011.

3.4. Hybrid Electric Bus Characteristics

An advantage of a hybrid-electric bus over a conventional bus is theoretically better fuel economy and lower exhaust emissions.

Capital, maintenance and operation cost: Hybrid buses can cost up to $500.000, a significant increase over a standard diesel city bus, the cost of which is closer to $300.000 [12]. The major cost associated with hybrid buses is battery replacement, as batteries today are not expected to last the 12-year life of a city bus.

Fuel and maintenance (operating) cost savings over the life of the bus are expected to help recover the higher initial (capital) cost. Specifically, operating cost savings are expected through the following features: increased fuel economy; extended brake life; no transmission to service; less moving parts; less engine wear and less expensive engine.

Emissions and fuel economy: With regard to emissions, clearly hybrids are not providing the zero emission. Nevertheless, some testing on hybrid buses has demonstrated that hybrids offer emissions benefits that are comparable to or better than clean diesel and CNG buses. There are four primary sources of efficiency and emissions reduction found in HEBs [13]: smaller engine size, regenerative braking, power-on-demand, and constant engine speeds and power output.

By adding an electric motor a hybrid electric bus can be equipped with a smaller, more efficient combustion engine.

Regenerative braking recovers energy normally lost as heat during braking, and stores it in the batteries for later use by the electric motor.

Another feature that saves energy and reduces emissions in HEBs is the ability to temporarily shut off the combustion engine during idle or coasting modes.

In a hybrid application, the bus can be designed to use its diesel engine only at the engine's optimum power output and engine speed range.

Barriers to widespread application: The hybrid vehicles are still a "work-in-progress". Many studies noted that hybrid buses still require improvements in three technology areas: energy storage, electrically-driven accessories and on-board diagnostics.

3.5.Comparative Analysis Between The Different Buses

An overall comparison of buses with different technologies (diesel, CNG and hybrid electric) have been realized in the COMPRO project and the results are presented in the report "Cost /effectiveness analysis of the selected technologies" [14]. A comparison of buses is based on several parameters, technological, financial, environmental and planning-based, such as reliability, eployment flexibility, fuel price, range, exhaust gas emissions, noise, extra infrastructual needs. The considerations made above are summarized in Table 3. Advantages of each compared technology are featured in green, disadvantages in red.

Another comparison is based on a West Virginia University's (WVU) study of city bus life cycle cost (LCC) [15, 16]. It covers the folowing bus types: diesel buses using ultra low sulfur diesel (ULSD), compressed natural gas (CNG) buses, and hybrid electric buses. LCC factors included capital costs (bus procurement, infrastructure, and emissions equipment) and operation costs (fuel, propulsion-related system maintenance, facility maintenance, and battery replacement) available from the literature.

A bus 12-year life cycle cost (LCC) analysis for a fleet of size of 100 buses was performed based on information available in the literature, manufacturers' specifications, and fuel economy data gathered by WVU [15, 16]. Only technology-dependent factors relevant to bus propulsion were considered; driver and management cost were excluded. Buses were assumed to operate at a national average speed of 12.7 mph (20.48 km/h), to travel for 32.007 mile (about 51.500 km) per year, and to seat 40 passengers for the purposes of calculation.

Table 3. The comparison between the different technologies of city buses.

Diesel Bus	CNG Bus	Hybrid Electric Bus
• Experienced	• New fuel filling station	• High range +
• High range	• Reduced range	• Low emission
• Pollution (PM + NO_x)	• Low emission / no PM	• Reduced consumption
• Noise	• Low fuel costs	• Reduced CO_2
• Increasing fuel costs	• Vehicle costs (+30.000 €)	• Reduced noise
• Dependence on mineral oil	• Dependence on natural gas	• Vehicle costs (+150.000 €)
		• Not yet experienced

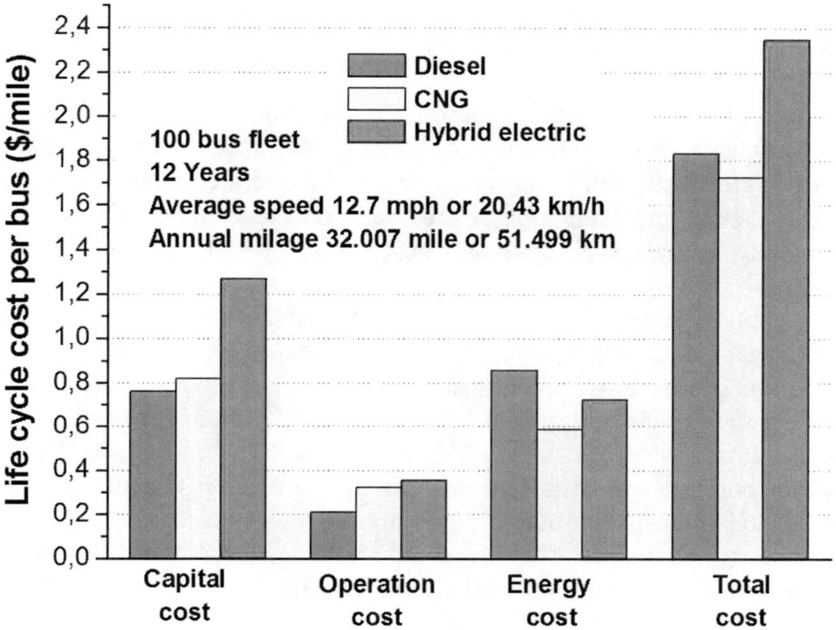

Figure 7. Comparison of life cycle cost between the different bus technologies.

Capital costs for vehicle procurement includes refueling station (CNG bus), depot modification, and emissions reduction equipment (diesel bus). Infrastructure costs for CNG bus technology include two costs: for depot modification and for refueling stations. Operational costs include compression electricity (CNG bus), facility maintenance, propulsion-related system maintenance, battery replacement (hybrid bus), and fuel consumption. Warranty was not considered. Fuel costs were calculated from the product of national annual average mileage, estimated fuel economy, and predicted fuel price. All prices were in 2008 dollars and CNG price data were all converted to the base of diesel (energy) equivalent.

Figure 7 representing total LCC was created for the capital and operation costs (without fuel consumption), and energy cost, per bus per mile.

The capital cost hybrid buses was slightly higher than CNG and diesel buses. However, operation cost analysis was similar for all bus types. Although hybrid buses offered the best fuel economy, this was offset by the battery replacement cost. Generally, the LCC total cost showed that diesel buses are still the most economic technology.

The West Virginia University's report [15] presents estimates for 2007 city bus regulated and greenhouse gas emissions. Tailpipe emissions (particulate matter (PM), nitrogen oxides (NO_x) and greenhouse gas (CO_2)) and fuel economy estimations were based on recent emissions and fuel economy studies, and adjusted with best engineering approach.

For simpler presentation of emissions and fuel economy by the three typical bus fleets (diesel, CNG and hybrid buses), the results given in the WVU study are appropriately processed and presented inFigure 8 [17]. Comparative values in Figure 8 represent average values for three typical driving cycles (MAN, OCTA, and CBD cycles). As can be seen from Figure 8 hybrid buses were attractive in offering emissions advantages. The estimation showed that hybrid buses offered lower tailpipe CO_2, NO_x and PM than the diesel and CNG buses. Figure 8 shows that the diesel hybrid bus fuel economy is better than the diesel bus fuel economy about 19%.

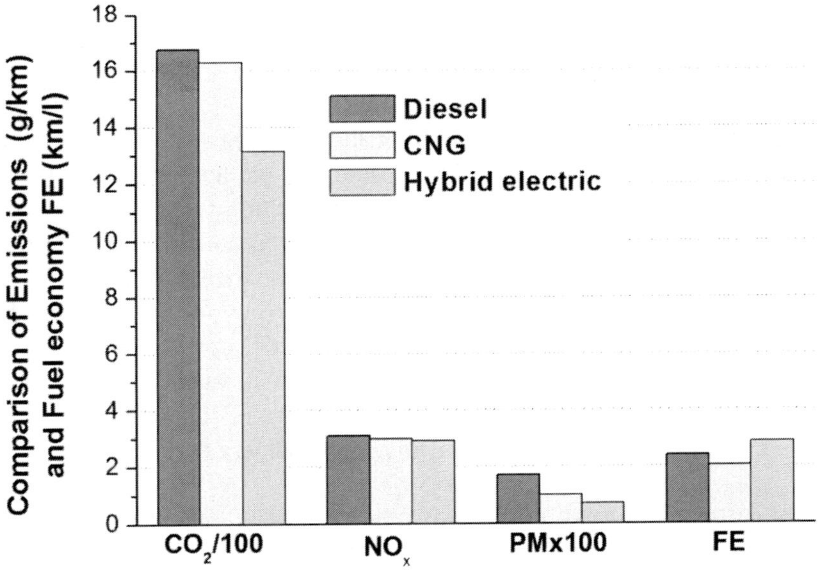

Figure 8. Comparison between the different bus technologies.

3.6. Hybrid Electric Bus Solutions

The use of hybrid technology has become a popular issue in recent years. Hybrid solutions are principally available for all main propulsion types, thus with Diesel, CNG, fuel cells etc. Table 4 [18,19] and Table 5 [20] show some typical hybrid electric bus solutions of major bus manufacturers with IC engine.

Besides the mentioned, other manufacturers of hybrid buses are: Gillig, ISE Corporation (North America); Scania, Irisbus Iveco, Van Hool, VDL Bus & Coach, Hess AG (Europe); Tata Motors, Toyota-Hino, Hyundai Motor Company, Mitsubishi Fuso (Asia).

4. FUEL CELL BUSES

Fuel cells for bus applications have generated an enormous amount of attention over the last several years, as they offer the promise of a clean, efficient transportation system no longer dependent on fossil fuels. The fuel cell has a life expectancy about one-half that of an internal combustion engine. Thus, consumers would have to replace the fuel cell twice in order to achieve a vehicle operating lifetime equivalent to that of a traditional engine.

Fuel cell buses (FCBs) require a substantial new infrastructure, support, and training requirements will depend on what type of fuel is used for the fuel cells. Most demonstrations and available buses use pure hydrogen stored in compressed gas form.

4.1. Fuel Cell Bus Configurations

All fuel cell vehicle concepts use electric motors to power the wheels, typically accomplished through the combination of an electric battery storage system and an on-board hydrogen fuel cell. Depending on the degree of hybridization, the battery may provide pure “plug-in” electricity to drive the vehicle some distance. The battery system would be complemented by a hydrogen storage system and a fuel cell, with the goal of extending the driving range to 300 miles (483 km). Early fuel cell bus designs involved an electric drive train, where a fuel cell generates electricity which is directly supplied to an electric motor. A new generation of FCBs is based on the hybrid concept. Figure 9 shows typical arrangement of the components in FCB [21]. Key system components are: fuel cell system, energy storage system, hydrogen storage system, wheel drive, cooling system, and auxiliaries.

Table 4. Hybrid electric buses manufactured in North America.

Hybrid bus	Orion VII Hybrid electric bus	New Flyer Hybrid-Electric bus
Image of bus	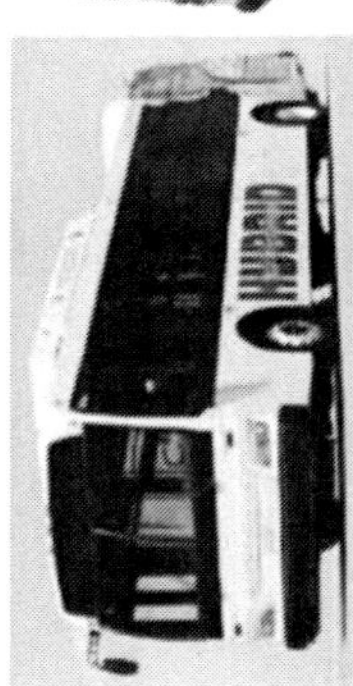	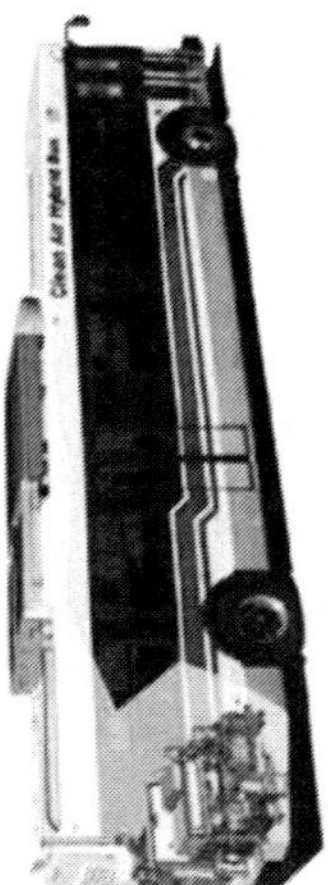
Type of hybrid drive	BAE Systems HybriDrive™ Serial hybrid	ISE ThunderVolt® TB40-HD Serial hybrid
Engine	Cummins ISB, ULSD, 5.9-litre, 194 kW, with a 120 kW traction generator	
Electric motor/generator	AC induction motor, Rated Power 250 hp continuous (320 hp peak)	Dual AC Induction motors, Rated Power 170 kW, Peak Power 300 kW
Energy Storage	Lithium-ion battery	Ultra capacitors
Bus Characteristics	Improve fuel economy 30% and reduce emissions: 90% PM, 40% NO_x, 30% CO_2	Reduce emissions: 25% PM, 32% NO_x and lower fuel consumption and CO2 emission

Table 5. Hybrid electric buses manufactured in Europe.

Hybrid bus	MAN Lion's City Hybrid	MERCEDES-BENZ Citaro G BlueTec Hybrid
Image of bus		
Type of hybrid drive	Siemens, Serial hybrid	Serial hybrid
Engine	MAN D0836 LOH CR, EEV, 6.9 litre, 191 kW /260 hp	OM 924 LA, 4.8 litre 160 kW/218 hp
Electric motor/ generator	2x75 kW Siemens asynchronous electric motors. Synchronized generator with an output power of 150 kW	Four electric wheel hub motors on the central and rear axles of the bus, output of 320 kW (4x80 kW). Generator output 160 kW
Energy Storage	Ultra capacitors	Lithium-ion battery

Bus Characteristics	Up to 30 percent lower fuel consumption and CO_2 emission	Diesel consumption and CO_2 emissions reduced by up to 30 %
Hybrid bus	VOLVO 7700 Hybrid	SOLARIS Urbino 18 Hybrid
Image of bus		
Type of hybrid drive	Integrated Starter Alternator Motor (I-SAM) , Parallel hybrid	Voith DIWAhybrid, Parallel hybrid
Engine	New Volvo D5E, 5,0 litre capacity, rated at 210 hp	Cummins ISB6.7 250H engine, 181 kW (246 hp), 6.7 litre capacity
Electric motor/ generator	AC permanent magnet motor, power rating of 160 hp and a continuous rating of 90 hp	The motor provides 85 kW of power continually with a maximum output of 150 kW
Energy Storage	Nickel- Metal-Hydride battery	Ultra capacitors
Bus Characteristics	Fuel savings up to 35 % Lower exhaust emissions	Average fuel saving up to 16%

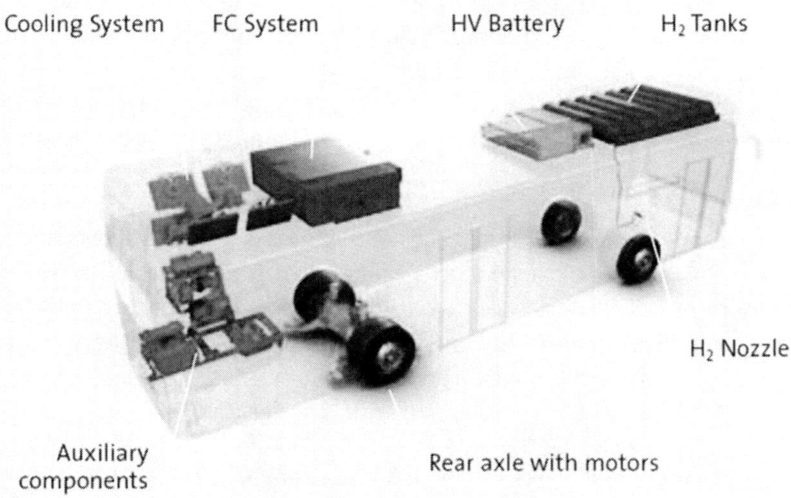

Figure 9. Key components of the fuel cell hybrid bus.

Fuel cell system: A fuel cell is an energy conversion technology that allows the energy stored in hydrogen to be converted back into electrical energy for end use. In a fuel cell vehicle, an electric drive system, which consists of a traction inverter, electric motor and transaxle, converts the electricity generated by the fuel cell system to traction power to move a bus. The fuel cell system and additional aggregates are usually located on top of the roof of the bus.

Fuel cells are classified by their electrolyte and operational characteristics. For application in vehicles mostly used are the Polymer Electrolyte Membrane (PEM) fuel cells. They are lightweight and have a low operating temperature. PEM fuel cells operate on hydrogen and oxygen from air.

Alkaline fuel cells (AFCs) are made by one of the most mature fuel cell technologies. AFCs have a combined electricity and heat efficiency of 60 percent efficient.

A newer cell technology is the Direct Methanol Fuel Cell (DMFC). The DMFC uses pure methanol mixed with steam. Liquid methanol has a higher energy density than hydrogen, and the existing infrastructure for transport and supply can be utilized.

There are some major fuel cell manufacturers supplying fuel cell power plants for heavy-duty applications: Ballard Power Systems and Hydrogenics (Canada), United Technologies Corporation (UTC) Fuel Cells, Enova Systems (USA), Shanghai Shen-Li High Tech Co. Ltd. (China), Siemens and Proton Motor Fuel Cell GmbH, (Germany), Toyota (Japan), Hyundai Motor Co. (South Korea).

Energy storage system: Energy storage systems are generally based on battery packs and/or ultracapacitors (generally up to 100 kW). Maximum power output and storage capacity vary depending on hybrid architecture. Lithium-ion battery technology is the most appropriate of energy storage technology for use in the buses. The batteries are usually located on top of the roof of the bus. FCBs are equipped with regenerative braking.

Hydrogen storage system: Gaseous hydrogen serves as the fuel. It is stored in compressed gas tanks, the number of which is decisive for the maximum range but also confines passenger capacity. The hydrogen storage system has been downsized as a result of the improved efficiency of the drive train. This has led to the reduction in the overall weight of the bus. The cylinders to storage hydrogen on board operate at an increased pressure of 350 bars.

Wheel drive: The electric motor can be either a single main motor or hub–mounted (where the motor is designed within the wheel). The bus may be equipped with a central traction system which will be located at the left hand side in the rear of the bus. The rear axle has 2 wheel hub motors and has been specifically developed to match the required speeds, load capabilities and energy efficiency. It also serves as a generator for energy regeneration during braking.

Cooling system: While running hydrogen through a fuel cell, water is of course being produced. Some of it becomes steam and leaves the system quite easily, as seen at the steam vent at the back of the bus. Yet, because PEM cells are sensitive to high heat the cell stacks must be cooled down. Therefore the byproduct from producing the electricity will always partially turn into liquid water that can accumulate in the stack and slow down the process. This can happen during idle times or at full speed. Therefore all PEM cells need a mechanism that clears the stacks once in a while or else the electricity production will be slowed down. The majority of the stack manufacturers use liquid cooled systems, with radiators to dissipate heat.

Auxiliaries: The auxiliary components in the FCB may be driven electrically. This means that they operate only on demand and are not driven continuously. This solution is typical for FCBs, which are based on the hybrid concept. This will result in a higher efficiency and lower maintenance of the components.

4.2. Fuel Cell Bus Characteristics

Fuel cells offer a number of potential benefits that make them appealing for transport use such as greater efficiency, quiet and smooth operation, and, if pure hydrogen is used on board the bus, zero emissions in operation and extended brake life. Infrastructure, buses, fuel, and maintenance costs associated with hydrogen fuel cells are currently prohibitively expensive. The cost of facilities has ranged from several hundred thousand dollars up to $4,4 million for a maintenance facility, fueling station, and bus wash [22]. Currently, fuel cells for buses are not a commercial product. The existing fuel cell buses are prototypes, manufactured in fairly small numbers. Fuel cell buses can cost $1 to $3 million (or more) since they are hand-built prototypes utilizing a pre-commercial technology.

The hydrogen fuel itself is also currently very expensive. Costs range depending on the method of hydrogen production. One of the major constraints for use of the fuel cell buses is the refueling time for hydrogen buses. Filling over 30kg of hydrogen in less than 5 minutes is not currently feasible without pre-cooling the hydrogen (as the temperature increase at these high fill rates would damage the hydrogen tanks) [23].

Some comparisons with different bus technologies (diesel, hybrid diesel-hybrid and fuel cell), in terms of CO_2 emissions per km traveled, have shown significant benefits of FCBs [23]. CO_2 emissions of fuel cell buses ranging from 0 to 1,8 kg/km. Zero emissions are related to renewable hydrogen and electricity. Emissions of diesel hybrid buses are 0,69 to 1,2 kg/km, but emission of diesel buses are 1,05 to 1,5 kg/km.

4.3. Fuel Cell Bus Projects

The introduction of new types of buses in urban public transport is sometimes a challenging process that includes testing, demonstration and limited production with a tendency to increase the number of vehicles. Fuel cell-powered buses continue to be demonstrated in public transport service at various locations around the world. Many demonstration projects have been launched in the last 10 years in various stages of implementation. Many have been completed, and some of them are still active. An overview of mainly fuel cell city bus development projects is given below [21]:

The HyFLEET:CUTE project has involved the operation of 47 hydrogen powered buses in regular public transport service in 10 cities on three continents. The Project started in 2006 and concluded at the end of 2009. HyFLEET:CUTE was co-funded by the European Commission and 31 Industry partners through the Commission's 6th Framework Programme.

ECTOS (Ecological City Transport System) was an initiative to test three Citaro fuel cell buses in Reykjavik, Iceland. The project was financially supported by the European Commission. The project started at mid-2003 and now the buses are decommissioned.

STEP (Sustainable Transport Energy Project) is a project to explore the use of alternative transport energies in Perth, Australia. It includes the operation three Citaro fuel cell buses. It is funded by the government of Western Australia with support from the Australian government. The project started at mid-2004 and now the buses are decommissioned.

CHIC (Clean Hydrogen In European Cities) is the active project which involves integrating 26 FC buses in daily public transport operations and bus routes in five locations across Europe – Aargau (Switzerland), Bolzano/Bozen (Italy), London (UK), Milan (Italy), and Oslo (Norway). The CHIC project is supported by the European Union and has 25 partners from across Europe, which includes industrial partners for vehicle supply and refueling infrastructure. The project will guide additional regions in Europe in their first fleet application of fuel cell hybrid buses in public transport from 2012 onwards [24]. The buses in the CHIC project will be supplied by three different manufacturers - Mercedes-Benz (Germany), Van Hool (Belgium) and Wrightbus (UK).

"NaBuZ demo" (Sustainable Bus System of the Future – Demonstration) is German-funded project [25] in which the Hamburger Hochbahn AG produced four Mercedes-Benz Citaro FuelCELL Hybrid buses. Since 2011, the first Citaro FuelCELL Hybrid bus is involved into service.

The National Renewable Energy Laboratory (NREL) has recently published a status report documenting progress and accomplishments from demonstrations of fuel cell city buses in the United States. The report describes the status and challenges of fuel cell propulsion for transport and summarizes the introduction of fuel cell city buses in North America. Three major programs are [26]:

1) *Federal Transit Administration's (FTA) National Fuel Cell Bus Program (NFCBP)* includes developing the 11 new buses, expanding the fuel cell manufacturers beyond Ballard and UTC Power to include Hydrogenics and Nuvera, and exploring multiple bus sizes and hybrid propulsion designs.

2) *Zero Emission Bay Area (ZEBA) Group Demonstration* includes 12 next-generation fuel cell city buses with redesigned Van Hool chassis, the newest UTC Power fuel cell power system, and fully integrated hybrid propulsion system.

3) *BC Transit Fuel Cell Bus Demonstration* includes 20 fuel cell buses in Whistler, Canada. The buses are from New Flyer, use Ballard fuel cells, and have hybrid propulsion from ISE/Siemens.

Other projects: In China more than 50 FCBs were used during the Asian Games in 2010 and the Olympics Games in 2008. In Japan, Toyota-Hino has launched several dozen fuel cell buses in the period since 2003. In Korea, a Hyundai fuel cell bus has operated since 2006 [27].

4.4. New Generation Of Fuel Cell Buses

Mercedes-Benz fuel cell hybrid buses: Mercedes-Benz has promoted the use of fuel-cell hybrid buses around the world over the past nine years. Since 2003, a total of 36 Mercedes-Benz Citaro buses equipped with second-generation fuel cells were driven a combined total of more than 2,2 million kilometres in a total of approximately 140.000 hours of operation [28].

Production of the second generation of Mercedes fuel-cell hybrid buses started in November 2010 under the CHIC project. Compared with the fuel cell buses which were tested in Hamburg in 2003, the new Citaro FuelCELL Hybrid, Figure 10, provides several significant new features [25]: hybridization with energy recovery and storage in lithium-ion batteries, powerful electric motors with 120 kW of continuous output in the wheel hubs, electrified power take-off units and further enhanced fuel cells. These should achieve an extended service life of at least six years or 12.000 operating hours.

New additions are the lithium-ion batteries which for example store recovered energy. With the power stored there the new Citaro FuelCELL Hybrid can drive several kilometres on battery operation alone. In general, the design of the new FuelCELL buses is largely the same as that of the Mercedes-Benz BlueTec Hybrid buses that run in regular service; these buses also get electrical energy from a diesel generator. Thanks to improved fuel cell components and hybridization with lithium-ion batteries the new Citaro FuelCELL Hybrid saves on almost 50 % in hydrogen usage compared with the preceding generation. Overall fuel cell system efficiency has also been improved. The fuel cell bus has a range of around 250 kilometres.

Figure 10. The new Citaro FuelCELL hybrid bus.

Van Hool fuel cell hybrid buses: VAN HOOL (Belgium) is the largest independent manufacturer of integral buses and coaches in Western Europe. More than 80% of the company's production is exported: two thirds stay in Europe, the remainder goes to America, Africa and Asia. In a joint effort with UTC Power (United Technologies Corporation), a supplier of fuel cell systems, Van Hool developed fuel cell buses for the European and U.S. markets. Siemens supplied the twin AC induction electric motor, 85 kW each, converters, and adapted traction software.

Figure 11. The new Van Hool fuel cell bus.

Within the project ZEBA demonstration includes 12 new generation fuel cell hybrid buses and two new hydrogen fueling stations [29]. The buses are 12 m, low floor buses built by Van Hool with a hybrid electric propulsion system that includes a UTC Power fuel cell power system (120 kW) and an advanced lithium-ion battery (rated energy: 17.4 kWh and rated power: 76 to 125 kW). Eight Dynetek, type 3 cylinders, 350 bars, are mounted on the roof. The new buses, Figure 11, feature significant improvements over two previous generations of fuel cell buses that were demonstrated in California, Connecticut, and Belgium. Improvements include a redesigned Van Hool chassis that is lighter in weight, shorter in height, and has a lower center of gravity for improved weight distribution. The bus has a top speed of 55 mph (88 km/h). The bus purchase cost is about $2,5 million.

4.5. Operating Cost and Fuel Economy Of Fuel Cell Buses

Many transport operators continue to aid the FCB industry in developing and optimizing advanced transportation technologies. These in-service demonstration programs are necessary to validate the performance of the current generation of fuel cell systems and to determine issues that require resolution.

By the end of June 2011, nine of the twelve new Van Hool fuel cell buses had been delivered and seven of those were in-service. The buses have accumulated more than 80.000 miles (128.000 km) and a total of 7.653 hours on the fuel cell systems. The results presented here are early/preliminary information from the first five fuel cell buses that have been placed into service at AC Transit [29].

Table 6. presents the comparative test results for the Fuel cell and Diesel buses during the evaluating period.

Table 6. Operating costs of different buses per kilometre.

	Fuel Cell	Diesel
Fuel Cost ($/km)	0.96	0.42
Total Maintenance Cost ($/km)	0.94	0.41
Total Operating Cost ($/km)	**1.87**	**0.83**

During 2011, NREL completed data collection and analysis on new generation FCB demonstrations at three transport operators in the United States: SunLine California, CTTRANSIT Connecticut, and AC Transit California [30]. The current-generation FCBs in service at AC Transit, CTTRANSIT, and SunLine were all of the same basic design: Van Hool (12m) buses with ISE Corp. hybrid-electric drives and UTC Power fuel cell power systems.

Table 7 shows the fuel economy of the buses at each location. Data are given in miles per diesel gallon equivalent and in km/litre (1mile per Gallon =0.425 km/litre).

The FCBs at the three locations showed fuel economy improvement ranging from 48% to 133% when compared to diesel and CNG baseline buses. This table also illustrates the variability of the results from fleet to fleet. The results are affected by several factors, including duty-cycle characteristics (average number of stops, average speed, and idle time). Average speed for AC Transit –9,8 mph (15,8 km/h); CTTRANSIT –6,5 mph (10,4 km/h), SunLine –13 mph (20,9 km/h) [30].

Table 7. Fuel economy of the fuel cell buses at different locations.

	FC bus		**Diesel bus**		**CNG bus**	
	mile/gallon	km/litre	mile/gallon	km/litre	mile/gallon	km/litre
AC Transit	6,8	2,9	4,2	1,8	-	
CTTRANSIT	5,5	2,3	3,7	1,6	-	
SunLine	8,0	3,4	-	-	3,5	1,5

Data about NO_x and PM emissions (per year) presented by Van Hool in its promotional materials for three different fleets of buses, are shown in Table 8 [31]. The data are calculated for 50.000 km per year, average speed 20 km/h, and power consumption 50 kW/h.

Equivalent emissions reduction potential of 100 hybrid fuel cell buses gives a CO_2 reduction equal to the uptake of 3.100 acres of forest and a NO_x reduction equal to 10 km of 4 lanes of cars [31]. The presented results show all the environmental advantages of the buses with fuel cell technologies.

Table 8. Comparative characteristics of NOx and PM emissions for different buses.

	NOx (per year)	PM (per year)
100 Diesel Euro III buses	62,5 tons	1,25 tons
100 CNG buses	25 tons	0,25 tons
100 Hybrid fuel cell buses	zero	zero

5. BATTERY ELECTRIC BUSES

Electric vehicles are a promising technology drastically reducing the environmental impact of road transport. At the same time, EVs are still far from proven technology. Reality is such that battery technology is simply not the whole answer. This is because (especially for large buses) batteries do not carry enough energy to power the bus for a full day.

According to IDTechEx report [32], "Electric Vehicles 2010-2020" it is estimated that worldwide there are about 480.000 electric buses, mostly small ones - with about 135.000 being bought each year as the fleets grow. Although only 12% of these new buses are electric, it is expected that by 2020 investment in the purchase of these buses will be measured in million of dollars. The main increase in electric buses will be running the free versions to be used without appropriate infrastructure along the route.

5.1. Full- Size Battery Electric Buses

Proterra's Battery Electric Bus: Proterra is the pioneering innovator and manufacturer of clean commercial transport solutions for city buses. Three Proterra's zero emission fast-charge battery electric buses, EcoRide BE35, Figure 12, have been put in service in 2010 [33].

The buses feature Proterra's revolutionary clean-transport technology, including the Proterra TerraVolt Energy Storage System, which allows a full battery recharge in less than 10 minutes. Additional bus features include:

- Flexible ProDrive and vehicle control system that can operate in battery electric mode or with any small APU to extend vehicle range when needed;
- Regenerative braking system, allowing the recapture over 90% of the vehicle's kinetic energy available during braking;
- Light-weight composite body resulting in 25% reduction in weight, significantly lower maintenance costs and 40% longer life than traditional diesel buses.

Figure 12. Proterra's battery electric bus.

Proterra's EcoRide BE35 battery powered buses can operate on standard routes for up to three hours—a range of 30-40 miles (48 to 64 km)—and after that, require just 10 minutes of charging to get back on the road. The buses can accommodate as many as 68 passengers and according to Proterra, will provide $300.000 in savings over the course of their lifetime thanks to lower fuel and transportation costs.

BYD electric bus: China's Build Your-Dream (BYD) hybrid buses, Figure 13, were showcased during the 2008 Beijing Olympics. BYD has developed and is currently marketing a lithium-ion battery - powered hybrid-electric bus and all electric buses.

BYD, manufacturer of the first long-range (>300 km) all-electric bus (eBUS-12), has been selected as the sole eBUS provider for the 2011 International Universiade Games held in Shenzhen, China. At the core of the eBUS technology is BYD's in-wheel motor drive system and the Iron Phosphate battery technology. The eBUS also integrates BYD solar panels on the bus roof, converting solar energy to electricity which is stored in the batteries and can completely offset the eBUS air-conditioning load (extending the range on sunny days) [34].

Figure 13. BYD eBUS-12 electric bus.

BYD has signed a Letter of Intent with the city of Frankfurt, Germany to introduce BYD's all-electric, long-range, eBUS. BYD will supply three all-electric buses eBUS-12, two DC charging stations and technical support in the first quarter of 2012.

BYD is set to trial a full-size, all-electric bus in the Danish capital Copenhagen. Two K9 buses will initially be deployed on ordinary passenger routes from the second half of 2012.

The City of Windsor, Ontario has signed a letter of intent to purchase up to 10 BYD fully electric buses for the community's transport services in 2012. It will become the first City in North America to launch long-range, all-electric buses. BYD has delivered over 300 all-electric buses worldwide and claims orders for over 1.300 more in 2012, making it the largest electric bus manufacturer in the world.

e-Traction buses: Netherlands' company e-Traction (European integrator of low-floor fleet buses), delivered in the year 2010 the first of two e-Busz electric drive buses to Rotterdam's public transportation authority. The e-Busz is a VDL Bus & Coach Citea CLF bus converted with the third generation of the e-Traction system [35].

e-Traction specializes in development of TheWheel as a direct-drive in-wheel motor system with integrated power electronics and fluid cooling. TheWheel is designed to deliver very high torque at low revolution. Since 2001, e-Traction is continuously developing in-wheel direct-drive motors for applications ranging from 400Nm to 10.000Nm per wheel. The vehicles with TheWheel save up to 40% traction energy and are 50% more fuel efficient compared to the standard diesel equipped bus. The e-Buzs is a "battery dominant" hybrid bus. This means that it has the ability to run on battery only, with the diesel generator turned completely off. The diesel unit (with diesel generator) can be replaced and, importantly, the bus returned to revenue service in roughly one hour.

In cooperation with e-Traction, Hybricon, a Swedish company converted two Volvo 7700 city buses to fully electric city bus with rapid charge technology. The buses with the name Arctic Whisper, Figure 14, are from November 2012 in public service in the city of Umeå, Sweden [36]. e-Traction and Hybricon removed the whole diesel driveline and replaced it with two e-Traction SM/500-3 wheel motors mounted on a rear axle construction. A 100 kWh Valence Li-Fe battery pack (for the purposes of the prototype) and pantographs for the charging station form the basic configuration, with a 50 kW diesel generator as the back-up system.

Figure 14. Arctic Whisper plug in hybrid bus.

The Arctic Whisper's bus is fast charged for 10 minutes at the end of its route to achieve nearly 100% all-electric operation but with the reliability of diesel. Without fast charging, the Arctic Whisper has an all-electric runtime of about 2-3 hours with the 100 kW batteries before the diesel generator needs to turn on.

Future plans include using different battery chemistries capable of faster charging and higher charging rates of over 200 kW as well as extending this architecture to 18 meter articulated buses.

SMG battery electric bus: The Seoul Metropolitan Government (SMG) has started commercial operation of full-size battery electric buses since 2010, Figure 15. The government has been working on a project to develop the buses with local technology after reaching an agreement with Hyundai Heavy Industries and Hankuk Fiber back in September, 2009. It now has a goal of putting 120.000 of the vehicles to use in the city by 2020 – this will account for 50% of all public transport vehicles [37].

The SMG electric buses are a low floor design, 11,00 m in length and can travel as far as 83 km on a single charge. Using high-speed battery chargers they can be fully charged in less than 30 minutes and have a maximum speed of 100 km/h. Four battery charges are being provided. They use high-capacity lithium-ion batteries and regenerative braking. To reduce their weight and help maximize the distance they can travel between charges, these buses make extensive use of carbon composite materials, instead of metal.

Figure 15. SMG battery electric bus.

Optare's battery-powered Versa: Optare's battery-powered Versa, Figure 16 - the UK's first commercial full-size battery bus which started on 1 April 2011 [38]. Battery-powered Versa buses are planned to be in service at peak times, providing a capacity increase on the 12-minute interval service. The buses can either be rapid charged, using a special charger, or slow charged. Travel de Courcey intends to re-charge the buses when they get to 30% of full charge. A return journey is around 5 miles (8 km). The Versa's testing is currently underway with the first bus to be delivered, pending the arrival of the other two, and their entry into service. They are not the first electric vehicles to be commercially produced by Optare. It has already delivered Solo EV models, also used on short distance shuttle services [38].

Tindo solar electric bus: The Adelaide City Council's electric solar bus, Figure 17, is the first in the world to be recharged using 100% solar energy. Recent advances in battery technology have helped the successful development of pure electric buses with a suitable range between recharges. The Tindo solar electric bus uses 11 Zebra battery modules, giving it unprecedented energy storage capacity and operational range [39]:

Some Tindo bus characteristic are: Length–10,42 m, Motor power peak – 160 kW, Motor power nominal - 36 kW, Speed - 76 km/h, Battery content - 261,8 kWh, Fast Charger Booster Power - 70 kW, Fuel costs - 50% lower than for a diesel bus, Range - 200 km between recharges under typical urban conditions.

Figure 16. Optare's battery-powered Versa.

Figure 17. Tindo solar electric bus.

5.2. Small Battery Electric Buses

Optare solo EV battery electric bus: Optare offers the Solo EV, Figure 18, fully electric buses available in lengths of 8,1m, 8,8m and 9,5m. Replacing the usual diesel engine is an all-new electric drive, featuring an Enova Systems P120 AC induction motor rated at 120 kW and powered by two banks of Valence Lithium Ion Phosphate batteries. The two packs work in parallel and provide 307 V with a total capacity of 80 kWh. [40].

Around 4.000 Optare Solo EV buses in service worldwide produce zero tailpipe emissions. The model demonstrated in Switzerland is based on a standard 27-seat, 8,8 metre Solo, but the technology can be used on other models in the Optare range with higher passenger capacities. The Solo EV has been designed to perform exactly like a standard diesel powered bus, except that it is smoother, quieter and cleaner. It is completely traffic compatible, with good acceleration and hill climbing capabilities and a top speed of up to 90 km/h. On a full charge it has a range of around 110-130 kilometres depending on load factors and topography.

Solaris Urbino electric bus: Polish bus manufacturer Solaris developed the midi Urbino electric bus,Figure 19. The innovative electric bus is based on the 8.9 m Urbino family midi bus. At the heart of its power system is a 120 kW four-pole asynchronous traction motor supplied by Vossloh Kiepe. Energy is stored in two batteries weighing 700 kg each. These liquid-cooled lithium-ion batteries have a rated voltage of 600 V and the capacity to store 120 kWh [41].

Figure 18. Optare Solo EV battery electric bus.

Figure 19. Solaris Urbino electric bus.

The battery capacity gives the Solaris Urbino electric a range of up to 100 km and a maximum speed of 50 km/h. Batteries are charged with a Walter plug-in connection. A full recharge from the 3x400 V, terminal takes as little as four hours. Even with the 1.400 kg traction batteries, the Solaris Urbino electric is only marginally heavier than its conventional counterparts thanks to the innovative lightweight construction employed.

6. ENERGY STORAGE SYSTEMS

Energy storage systems, usually batteries, are essential for electric drive vehicles. Batteries must have a high energy-storage capacity per unit weight and per unit cost. Because the battery is the most expensive component in most electric drive systems, reducing the cost of the battery is crucial to producing affordable electric drive vehicles.

The electrical energy storage units must be sized so that they store sufficient energy (kWh) and provide adequate peak power (kW) for the vehicle to have a specified acceleration performance and the capability to meet appropriate driving cycles. For those vehicle designs intended to have significant all-electric range, the energy storage unit must store sufficient energy to satisfy the range requirement in real-world driving. In addition, the energy storage unit must meet appropriate cycle and lifetime requirements. These requirements will vary significantly depending on the vehicle type (battery or fuel cell powered or hybrid electric).

There are many energy storage technology and battery chemistry and packaging options for electric drive buses.

6.1. Energy Storage Device Used In Some Electric Drive Buses

Based on the presentation of realized solutions of electric drive buses, in the previous items of this chapter, it may be concluded that in the energy storage devices the latest technology batteries and ultra capacitors are applied. Summary of main characteristics of energy storage devices, for each of the presented buses is given in Table 9.

As can be seen from Table 9, Li-ion batteries are prevailing in the realized bus solutions. In the latest bus solutions Zebra battery and Iron Phosphate battery technology are used. It is notable that the energy capacity of energy storage devices BEBs are significantly higher than that of HEBs and FCBs.

In a HEBs with an ICE that recharges the battery (where battery operates in charge sustaining mode), a lighter and smaller battery is employed.

Table 9. Characteristics of energy storage devices of several bus solutions.

	HYBRID ELECTRIC BUSES	
BUS	MAN Lion's City Hybrid	MERCEDES-BENZ Citaro G BlueTec Hybrid
Energy storage	Capacitors Energy content: approx. 0,5 kWh, Max. charging/discharging power: 200 kW, Voltage: 400-750 V	Lithium-ion battery Energy content:19,4 kWh, Maximum output of 240 kW and, Located on the roof
BUS	VOLVO 7700 Hybrid	SCANIA Ethanol hybrid bus
Energy storage	Nickel-Metal-Hydride battery Energy content: approx. 4,8 kWh, Weighting approximately 350 kg, Rated at 600 volts, Located on the roof	Supercapacitors Energy available: "/400 Wh, 4x125-Volt Maxwell BOOSTCAP® modules, air-cooled, Design life:10-15 years
BUS	SOLARIS Urbino 18 DIWA Hybrid	ORION VII hybrid electric bus

Energy storage	Super capacitors Energy content: 0,5 kWh, Maxwell, 5x125V, Weight 410 kg	Lithium-Ion battery Energy content: 32 kWh, Weight 364 kg, 6 year design life, Roof-mounted
	FUEL CELL BUSES	
BUS	Mercedes-Benz Citaro FuelCELL Hybrid bus	Van Hool Fuel Cell Hybrid Bus
Energy storage	Li-ion Battery Energy content: 26 kWh, Energy storage power 250 kW	Li-ion Battery Energy content: 17,4 kWh, Rated power: 76 to 125 kW
BUS	New Flyer Fuel Cell Bus	Van Hool Fuel Cell Hybrid Bus (UTC power)
Energy storage	Li-ion Battery Energy content: 47 kWh	NaNiCl (ZEBRA) battery Energy content: 53 kWh
	BATTERY ELECTRIC BUSES	
BUS	Optare Solo EV Battery Electric Bus	Solaris Urbino Electric Bus
Energy storage	Li-ion Battery Energy content: 80 kWh	Li-ion Battery Energy content: 120 kWh, Rated voltage of 600 V
BUS	BYD eBUS-12 Battery Electric Bus	Tindo Solar Electric Bus
Energy storage	Li Iron-Phosphate or "Fe" battery Energy content: 324 kWh Milage: 300 km on a single charge	Li-ion Battery Battery content: 261,8 kWh

A BEBs require a larger and heavier battery pack, to provide both high energy density and high energy storage capacity so as to maximize the range between recharges.

In the PHEBs one can expect smaller, intermediate-sized, battery packs capable of either charge-sustaining operation in the blended mode with an active ICE, or charge-depleting operation.

6.2. Current And Future Development Of Energy Storage Devices

A number of different battery technologies exist at present. The lead acid battery has been used to supply vehicle electricity for a number of decades. With the introduction of the first modern EVs in the 1980s, the need for more powerful batteries arose. Nickel-cadmium batteries were originally used, later replaced in hybrid vehicles by nickel-metal hydride batteries.

However, none of these battery technologies provide the energy density required for sufficient driving distance in pure electric mode.

Recently, apart from the mentioned, other energy storage devices are in the intensive growth and expansion, such as: Lithium-ion battery (Li-ion), Li-ion polymer battery, Sodium Nickel Chloride battery (NaNiCl), Lithium iron phosphate battery (LiFePO4), Zinc Air battery, and Supercapacitors.

Based on available analysis and current battery data, it appears that the current (2010) battery life should exceed seven years and may be around ten years for 'average' use. The most promising chemistries appear to involve silicon, sulfur and air (oxygen) and another important development is research into nanotechnologies. These trends have been widely recognized and a recent presentation by Limotive researchers showed the following battery technology roadmap, Figure 20 [41].

Silicon is an attractive anode material for lithium-ion batteries because it has about ten times the amount of energy that a conventional graphite-based anode can contain [42]. It also has a specific energy of 1.550 Wh/kg – about four times the energy of a conventional graphite-based anode. Furthermore, silicon is the second most abundant element on the planet and has a well-developed industrial infrastructure, making it a cheap material to commercialize with a cost comparable to graphite per unit of weight.

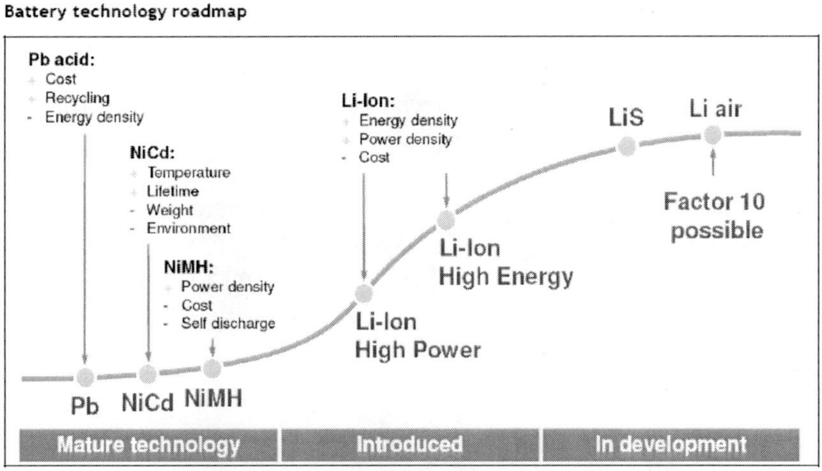

Figure 20. The battery technology roadmap.

The problem with silicon is that it is very brittle and when lithium-ions are transferred during charge and discharge cycles, the volume expands and contracts by 400% which can pulverise the silicon anodes after just the first cycle.

The Li-Ion technology will become more and more the dominant technology for electro mobility. The Li-Ion technology has not yet reached its full potential, further improvements are still possible. Further developments are needed to improve capacity and lifetime, reduce volume and costs (currently around €250-€500/kWh for NiMH and €700-€1.400/kWh for Li-ion), and to be safe and reliable [43].

Although few serious technical hurdles remain to prevent the market introduction of electric powered vehicles, battery technology is an integral part of these vehicles that still needs to be significantly improved. Both current and near-term battery technologies still have a number of issues that need to be addressed in order to improve overall vehicle cost and performance. These issues include [44]:

- *Battery storage capacity* – Batteries for EVs need to be designed to optimize their energy storage capacity, while batteries for PHEVs typically need to have higher power densities.

- *Battery duty (discharge) cycles* – Batteries for various electric powered vehicles have different duty cycles. Batteries may be subject to deep discharge cycles (in all electric mode) in PHEVs or frequent recharge cycles through regenerative braking in conventional HEVs. Batteries for EVs will be subjected to repeated deep discharge cycles without as many intermediate cycles. Current battery deep discharge durability will need to be significantly improved.

- *Durability, life expectancy, and other issues* – Batteries must improve in a number of other respects, including durability, life-expectancy, energy density, power density, temperature sensitivity, reductions in recharge time, and reductions in cost. Battery durability and life-expectancy are perhaps the biggest technical hurdles to commercial application in the near-term.

7. CONCLUSIONS

A significant part in the future reduction of consumption of fossil fuels and of the corresponding reduction of emissions of harmful gases will be played by the alternative propulsion systems and alternative fuels. The development of electric drive technologies intended for application in buses is expanding. However, there are many limitations which at this stage slow down these developments. Sustainability of alternative propulsion systems is dependent upon the degree of their technological development and a compromise between the opposed economical, ecological, and social factors [45].

A large number of hybrid buses in North America and Japan and their intense development in Europe over the past several years is a confirmation that their number in the near future will be permanently growing. Although the sale of these vehicles is relatively small, the high cost of fossil fuels and the costs of hybrid vehicles becoming more acceptable will accelerate their further development. The hybrid buses are expected to contribute to further reduction of CO_2 emissions, even though some manufacturers have reached the level of 30%. Further improvements in that direction will be dependent on the degree of hybridization of the propulsion system and electronic control which should contribute to the optimization of operation of ICE and hybrid system as a whole.

The experiences so far acquired, through the development of fuel cell buses and many demonstration projects around the world, are very positive. Some reports indicate that the performances of fuel cell buses in-service are above expectation. One of the many barriers to their wider use is the uncertainty of hydrogen supplies and high production costs of hydrogen. Other barriers are related to the security aspects of usage of these vehicles. Despite the insufficient performances of batteries, the next generation of fuel cell buses will be based on the hybrid concept and Li-ion batteries. Some predictions tell that fuel cells for buses will be commercially available within the next 10-15 years. At present, the high cost of buses is one of the greatest barriers to their commercialization.

Since 2010 increasing the number of battery driven vehicles and buses is evident, thus it can be expected that in the forthcoming times their number will continue increasing. However, barriers to their massive implementation will be the radius of movement, lack of infrastructure for recharging the batteries and, of course, high cost of the batteries and other power equipment (electric motors and control electronics).

Further challenges for electric drive buses will be the development of battery technologies and of other energy sources. Even though a considerable advancement has been made over the past several years by the development of Li-ion batteries, which have achieved energy density of 95-190 Wh/kg, there is still space for further advancements. The only battery chemistries that have a chance of achieving energy densities in the 1,000 Wh/kg range are rechargeable metal-air and other. Other non-chemical energy storage devices include super-capacitors that can reach very high specific power levels for a few seconds, but cannot hold a lot of energy.

The current generation of lithium-ion batteries typically uses a carbon-based anode and a metal oxide cathode. Research on next generation lithium batteries will continue the development of electrode and electrolyte materials and chemistries in order to increase the life and energy density of the battery while reducing size and weight. The most promising chemistries appear to involve silicon, sulphur and air (oxygen) and another important development is research into nanotechnologies.

ACRONYMS AND ABBREVIATIONS NOMENCLATURE

AC-Alternating Current

AFC-Alkaline Fuel Cell

APTA-American Public Transportation Association

APU-Auxiliary Power Unit

BEB-Battery Electric Bus

BEV-Battery Electrics Vehicle

BYD-Build Your-Dream

CARB-California Air Resources Board

CBD-Central Business District

CHIC-Clean Hydrogen in European Cities

CNG-Compressed Natural Gas

CO-Carbon Monoxide

CO_2-Carbon Dioxide

COMPRO -COMmon PROcurement of collective and public service transport clean vehicles

CUTE -Clean Urban Transport for Europe

DC -Direct Current

DMFC -Direct Methanol Fuel Cell

ECTOS-Ecological City TranspOrt System

EHPS-Eaton Hybrid Power System

EM/G-Electric Motor/Generator

EV-Electric Vehicle

FCB-Fuel Cell Bus

FCEV-Fuel Cell Electrics Vehicle

FTA-Federal Transit Administration

HEB-Hybrid Electric Bus

HEV-Hybrid Electrics Vehicle

ICE-Internal Combustion Engine

ISAM-Integrated Starter Alternator Motor

LCC-Life Cycle Cost

LiFePO4 Lithium Iron Phosphate

Li-ion-Lithium-ion

LNG-Liquid Natural Gas

MAN-Manhattan Cycle

NaNiCl-Sodium Nickel Chloride

NO_x -Nitrogen Oxides

NFCBP-National Fuel Cell Bus Program

NREL-National Renewable Energy Laboratory

OCTA-Orange County Transit Authority

PEM-Polymer Electrolyte Membrane

PHEB-Plug-in Hybrid Electric Bus

PHEV-Plug-In Hybrid Electrics Vehicle

PM-Particulate Matter

SMG-Seoul Metropolitan Government

STEP-Sustainable Transport energy for Perth

ULSD-Ultra Low Sulfur Diesel

UTC-United Technologies Corporation

WVU-West Virginia University

ZEBA-Zero Emission Bay Area

ACKNOWLEDGEMENTS

Financial support by Ministry of Education and Science Republic of Serbia (Projects TR 35041, TR 35042 and TR 35036) is gratefully acknowledged.

REFERENCES

1. C. Chan, 2007 The State of the Art of Electric, Hybrid, and Fuel Cell Vehicles Proceedings of the IEEE 95 4 704 718

2. Battery Hybrid and Fuel Cell Electric Vehicles are the keys to a sustainable mobility, AVERE, European Association for Battery Hybrid and Fuel Cell Electric Vehicles 2010 http://www.avere.org/www/ Images/files/about_ev/ Brochure.pdf accessed 2 November, 2011.

3. Hybrid Drives History of the vehicles with hybrid drive DAIMLER 2012 http://www.daimler.com/dccom/0-5-1200802-1-1400984-1-0-0-1201129-0-0-135-7165-0-0-0-0-0-0-0.html

4. New York City Hybrid Bus Overview 2012 http://ebookbrowse.com/04-compro-ws-hb-labouff-pdf-d296145507 accessed 15 March

5. About BAE Systems 2012 http://www.hybridrive.com/about-bae-systems. asp accessed 10 May

6. European bus fleets reorder Allison hybrid technology to reduce emissions and fuel consumption 2011 Allison Transmission http://www.propel-technology.com/Article__Europeanbusfleets reorderAllisonhybridtechnologytoreduceemissionsandfuelconsumption .aspx accessed 12 March

7. A Brecher, 2010 Assessment of Needs and Research Roadmaps for Rechargeable Energy Storage System Onboard Electric Drive Buses U.S. Department Of Transportation, Federal Transit Administration Report No. Fta-Tri-Ma-26-7125-2011.1 accessed 12 May 2012http:// ntl. bts.gov/lib/ 35000/35700/35796/DOT-VNTSC-FTA-11-01.pdf

8. ZF and ISE collaborate on parallel-electric hybrid drive systems, EOBUS 2009 http://www.eobus.com/news/447.htm accessed 12 January 2012

9. Eaton Powering Business Worldwide. 2012http://www.eaton.com /Eaton/ProductsServices/ProductsbyCategory/HybridPower/SystemsOv erview/ElectricHybrid/index.htm accessed 10 February

10. E. Jobson, 2010 Volvo 7700 Hybrid, The worlds most environmentally friendly bus Volvo Bus Corporationhttp://www.voev. ch/dcs/users/ 117/2_7_Volvo_Edward_Jobson.pdf accessed 20 December

11. DIWAhybrid, Voith Turbo GmbH & Co. KG, G 1977 e 2011-10, http://www.voithturbo.com/applications/vt-publications/downloads/1273_e_g_1977_e_diwahybrid_2011-09_screen.pdf

12. Hybrid Transit Buses, Cost and Maintance The Union of Concerned Scientists' Hybridcenter.org 2010 http://www.hybridcenter.org/hybrid-transit-buses.html accessed 10 February 2012

13. Hybrid Buses, Costs and Benefits Environmental and Energy Study Institute (EESI) 2012 Washingtonhttp://www. eesi.org/files/eesi_hybrid_bus_032007.pdf 21 April 2012

14. S Zamboni, A Normanno, 2008 D2.3 Cost/effectiveness analysis of the selected technologies (CNG and HYBRIDS) COMmon PROcurement of collective and public service transport clean vehicles, Bologna http://www.comproeu.org/doc/COMPRO_deliverable_D2.3_final.pdf (accessed 13 February 2010).

15. N. C Nigel, Z. W Feng, W Scott, W. L. Donald, 2007 Transit Bus Life Cycle Cost and Year 2007 Emissions Estimation U.S. Department of Transportation Federal Transit Administration FTA-WV-26-7004.2007.1, Final Report accessed 27 August 2010http://www. fta.dot.gov/documents/ WVU_FTA_LCC_Final_Report_07-23-2007.pdf

16. N. C Nigel, Z. W Feng, W. Scott, W. L. Donald, 2008 Additional Transit Bus Life Cycle Cost Scenarios Based on Current and Future Fuel Prices U.S. Department of Transportation Federal Transit Administration FTA-WV-26-7006.2008.1, Final Report accessed 27 August 2010http://www.fta.dot.gov/documents/WVU_FTA_LCC_Second_Report_11-03-2008.pdf

17. Z. Zivanovic, Z. Jovanovic, Z. Sakota, 2011 A Comparative Analysis of CNG and Hybrid Buses vs Diesel Buses 10th Anniversary International Conference on Accomplishments in Electrical and Mechanical Engineering and Information Technology, DEMI, Proceedings 607 612

18. DAIMLER Daimler Buses North America Orion VII Hybrid Product Overview 2012http://www.orionbus.com/Projects/c2c/channel/documents/845706_Orion_Hybrid_Transit_Presentation_Q1_2009.pdf accessed 21 April

19. Diesel Hybrid-Electric Drive System ISE ThunderVolt® Transit Bus Hybrid Diesel (TB-HD) 2009 http://www.isecorp.com/2009/wp-content/uploads/2009/02/diesel-hybrids-09-08-ebook.pdf (accessed 19 May).

20. Z. Zivanovic, Dj. Diligenski, Z. Sakota, 2009 The Application of Hybrid Drive Technology in City Buses XXII. JUMV International Automotive Conference Science & Motor Vehicles Belgrade Proceedings on CD 1 15

21. HYDROGEN TRANSPORTS. Bus Technology & Fuel for TODAY and for a Sustainable Future A Report on the Achievements and Learnings from, The HyFLEET:CUTE Project 2006 2010 http://hyfleetcute. com/data/HyFLEETCUTE_Brochure_Web.pdf accessed 30 August

22. B. Allison, C. Spencer, P. Hannah, P. Claire, R. Kayla, 2008 The Feasibility of Alternative Fuels and Technologies An Assessment of Addison County Transit Resources' Current and Future Options Middlebury Collegehttp://www.middlebury.edu/media/view/255373/original/BUSFINALREPORT.pdf accessed 27 December 2010

23. R Zaetta, B Madden, 2011 Hydrogen Fuel Cell Bus Technology State of the Art Review Project EC FCH-JU-2008-1 Grant Agreement Number 245133, Report Version 3.1 http://nexthylights.eu/Publications/Clean-3_D3-1_WP3_EE_State_of_the_Art_23rd-FEB-2011.pdf accessed 19 May 2012

24. CHIC In Brief 2012 http://chic-project.eu/about/background/chic-in-brief accessed 10 June

25. Hamburg putting its first Citaro FuelCELL Hybrid buses into service, Green Car Congress 2011http://www.greencarcongress.com/2011/08/hamburg-putting-its-first-citaro-fuelcell-hybrid-buses-into-service-hamburger-hochbahn-ag-has-acquired-four-mercedes-benz.html accessed 4 March 2012

26. L. Eudy, K. Chandler, C. Gikakis, 2009 Fuel Cell Buses in U.S. Transit Fleets: Current Status National Renewable Energy Laboratory Technical Report NREL/TP-560-46490

27. R Ahluwalia, X Wang, R Kumar, 2012 Fuel Cell Transit Buses Argonne National Laboratory, Argonne, ILhttp://www.ieafuelcell. com/documents/ Fuel_Cells_for_Buses_Jan_2012 accessed 10 June 2012

28. New Citaro Fuel Cell Bus Debuts in Hamburg 50% Lower Fuel Consumption Green Car Congress, 2009http://www.greencarcongress.com/2009/11/citaro-fc-20091116.html accessed 24 March 2011

29. C. Kevin, E. Leslie, 2011 Zero Emission Bay Area (ZEBA) Fuel Cell Bus Demonstration: First Results Report, Technical Report, NREL/TP-5600-52015 August http://www.nrel.gov/hydrogen/pdfs/52015.pdf Accessed 10 May 2012

30. E. Leslie, 2011 VII.5 Technology Validation: Fuel Cell Bus Evaluations FY Annual Progress Report, DOE Hydrogen and Fuel Cells Program 1036 1039 http://www.hydrogen.energy.gov/pdfs/progress11/vii_5_eudy_2011.pdf

31. Hybrid fuel cell buses 2012 Van Hool NV, Belgium http://www.vanhool.be/FRA/transport-public/hybride-pile-a ombustible/Resources/folderFuelCell. pdf Accessed 2 May

32. The future of electric buses 2011 Energy Harversting Journal http://www.energyharvestingjournal.com/articles/the-future-of-electric-buses-00002792.asp?sessionid=1 Accessed 2 November

33. J. Goldman, 2009 Fast Charge Battery Electric Transit Buses, Economically and Environmentally Sustainable, Simple and Safe Advanced Technology Proterra LLC, ARB ZEB Workshop http://www.arb.ca.gov/msprog/bus/zeb/meetings/0509workshops/proterra.pdf accessed 18 May 2012

34. Largest Electric Bus Fleet in World Launches- BYD Delivers First Orders 2011 CHINA4AUTO.comhttp://www.china4auto.com/ennews/html/Commercial_Vehicle/20110513/130526697818626.html accessed 5 May 2012

35. E-Traction- Company Profile 2012 E-Traction Europe B.V http://www.e-traction.eu/about-e-traction/our-company accessed 12 June

36. Project: Hybricon City Busses 2012 Umeå, Sweden, E-Traction Europe B.V http://www.e-traction.eu/projects/hybricon accessed 12 June

37. P. Lucas, 2010 Electric buses begin operation TheGreenCarWebsite.co.ukhttp://www.thegreencarwebsite.co.uk/blog/index.php/2010/12/28/electric-buses-begin-operation/ accessed 12 June, 2012

38. M. Holley, 2012 Battery Versa goes on trial Optare, an Ashok Leyland Company May 11http://www.optare.com/images/pdf/De%20Courcey %20Versa%20EV%201.pdf accessed 12 June 2012

39. Tindo 2012 The World's First Solar Electric Bus The Adelaide City Councilhttp://www.adelaidecitycouncil.com/assets/acc/Environment/e nergy/docs/tindo_fact_sheet.pdf accessed 12 June

40. Optare Introduces Battery-Electric Bus Green Car Congress 2009 http://www.greencarcongress.com/2009/03/optare-introduces-batteryelectric-bus.html accessed 12 June 2012

41. Urbino electric Characteristic Solaris Bus & Coach http://www. solarisbus.pl/en/electric.html accessed 12 May 2012

42. D. Gopalakrishnan, H. Essen, B. Kampman, M. Grünig, 2011 Impacts of Electric Vehicles-Deliverable 2 Assessment of electric vehicle and battery technology, Delft, CE Delft

43. ETSAP (Energy Technology Systems Analysis Program) 2012 TECHDS Energy Technology Briefs http://www.iea-etsap.org/web/ Demand. aspAccessed 11 April

44. Technology Roadmap 2011 Electric and plug-in hybrid electric vehicles International Energy Agency (IEA)http://www.iea.org /papers/2011/ EV_PHEV_Roadmap.pdf

45. Z. Nikolic, Z. Filipovic, Lj. Janjusevic, 2011 State of development of the electric and hybrid vehicles, energetic and ecological aspect of applications Industry 34 4 267 292

Chapter 8

Standardization Work for BEV and HEV Applications: Critical Appraisal of Recent Traction Battery Documents

Noshin Omar [1,2,*], Mohamed Daowd [1], Omar Hegazy [1], Grietus Mulder [3], Jean-Marc Timmermans [1], Thierry Coosemans[1], Peter Van den Bossche [2] and Joeri Van Mierlo [1]

[1]Vrije Universiteit Brussel, Pleinlaan 2, Brussel 1050, Belgium

[2]Erasmus University College, Nijverheidskaai 170, Brussel 1070, Belgium

[3]VITO, Unit of Energy Technology, Boeretang 200, Mol 2400, Belgium

ABSTRACT

The increased activity in the field of Battery Electric Vehicles (BEVs) and Hybrid Electric Vehicles (HEVs) have led to an increase in standardization work, performed by both world-wide organizations like the IEC or the ISO, as by regional and national bodies such as CEN, CENELEC, SAE or JEVA. The issues of these standards cover several topics: safety, performance and operational/dimension issues. This paper reports a brief overview of current standardization activities of lithium batteries based on IEC 62660-1/2 and ISO 12405-1/2. Furthermore, in this paper, a series of innovative test procedures for lithium-ion batteries are presented. Thanks to these tests, the general characteristics of a battery such as charge and discharge capabilities, power performances and life cycle can be determined. Then, a new approach for extracting the life cycle of a battery in function of depth of discharge has been developed.

KEYWORDS

lithium batteries; energy density; power density; performance tests; HPPC-test

1. INTRODUCTION

As the global economy begins to strain under the pressure of rising petroleum prices and environmental concerns, research has spurred the development of various types of Hybrid Electric Vehicles (HEVs), Battery Electric Vehicles (BEV) and Plug-in Electric Vehicles (PHEV). HEVs and PHEVs usually require more than one energy source to provide more efficient propulsion. The energy sources should satisfy the following basic operational requirements [1]:

- Sufficient power capabilities, so that the necessary power required for propulsion can be supplied to the motor in any reasonable driving condition;
- Quick charging time, in order to increase vehicle availability;
- Sufficient lifetime, both in terms of calendar life and number of charge/discharge cycles;
- Cost.
- Unfortunately, there is no single available battery on the market that satisfies all the criteria as stated above. One of the more serious problems with these vehicles is the present dependence on very expensive batteries such as nickel metal hydride (NiMH). At present, lead-acid actually is the only economically attractive battery, but it is ill-suited for HEV's because of the frequent cycling and the fact that the state of charge (SoC) must be held below about 70% to accept regenerative energy [2,3,4,5]. The Ragone plot in Figure 1 shows that lithium batteries have the best performance in terms of energy density

and power density. It should be noted however that the plot area represents several battery designs optimized for either power or energy.

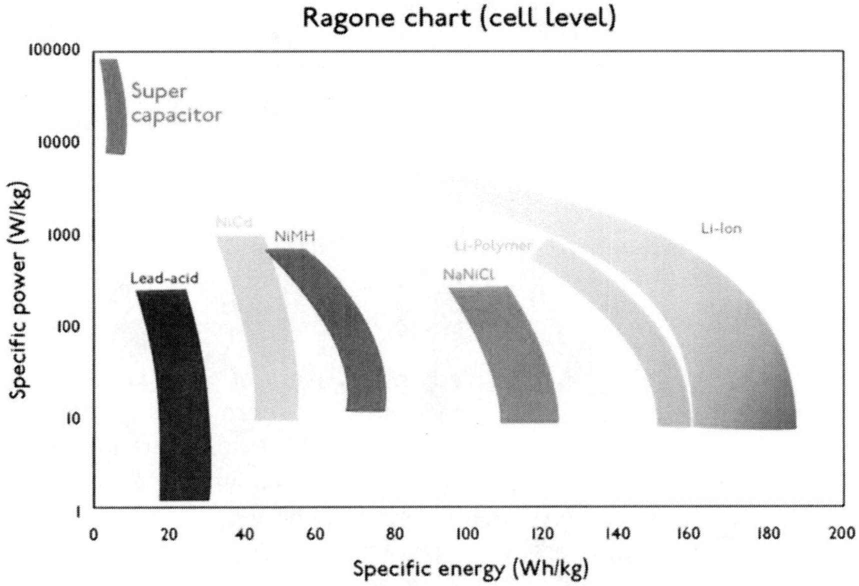

Figure 1. Rangone plot [3].

Lithium-ion batteries have shown their suitability in portable applications because of their high specific energy, lack of memory effect, and slow self-discharge when not in use [6]. In addition to consumer electronics, lithium-ion batteries are increasingly used in defence, automotive, and aerospace applications due to their superiority over any battery technology in terms of high energy density and high power capabilities. Several types of lithium technologies can be found on the market, such as Lithium Iron Phosphate (LFP), Lithium Nickel Cobalt Aluminum (NCA), Lithium Nickel Manganese Cobalt (NMC), Lithium Titante Oxide, *etc.* (Figure 2) [7,8]. All these batteries differ in capacity, energy, performances and potential safety hazards. In order to investigate the applicability of lithium-ion batteries, especially power and energy capabilities, in BEV and HEV applications; they shall to be subjected to specific test procedures.

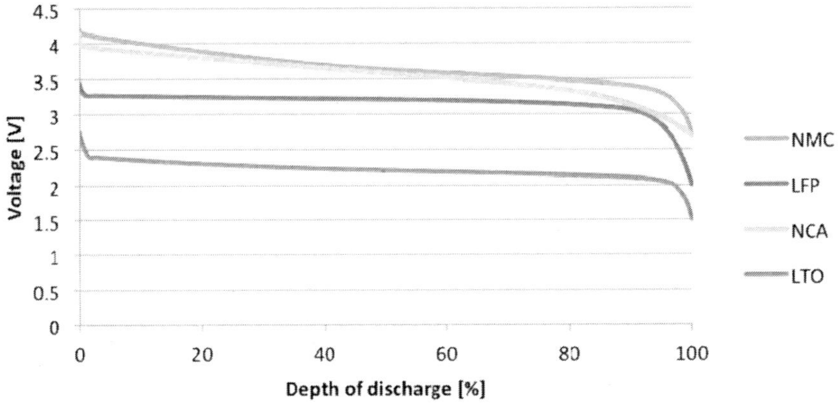

Figure 2. Voltage duration of various lithium-ion chemistries.

In the literature, one can find a number of papers regarding characterization of batteries for modeling and performances purposes. In [9,10,11,12,13,14,15,16,17,18,19] some test methodologies are introduced in order to obtain the general characteristics of lithium-ion batteries. However, the used test procedures are mainly based on constant current rates during discharge phase. These profiles are intended to investigate the discharge capabilities and to study the thermal behaviour and hysteresis curves of the examined batteries. In [20] Andre *et al.* used the sinusoidal load profiles for characterization of high power lithium-ion batteries by using electrochemical impedance spectroscopy. However, the current level of such measurement devices is limited to few amperes. In [21] Lee *et al.* have applied real load profiles for the evaluation of battery behaviour. In addition, in [22,23,24,25], asymmetrical and symmetrical load profiles have been implemented for emulating the battery voltage response during discharge phase. However, these mentioned profiles are only useful to investigate a well-defined problem. In order to be able to investigate and to compare the battery performances at different conditions, there is a need to use test procedures where all relevant battery performance characteristics can be derived. Moreover, these tests should be able for modeling issues, where emphasis is on extracting the model parameters. In the standard documents ISO 12405-1, ISO 12405-2 and IEC 62660-1/2 [26,27,28,29], some general test procedures in order to extract the power and energy densities of the batteries are defined. However, each test method is only suited to get one-performance characteristics. The characterization of all battery parameters is by this way time consuming and cumbersome, and is almost all at constant currents.

In this paper, the latest standardization works in the fields of lithium-ion batteries for BEV and HEV applications are extensively discussed and evaluated. The defined test procedures in the standards IEC 62660-1/2 and ISO 12405-1/2 have at first been analyzed based on experimental implementation of the test procedures in a battery test equipment. Then the tests have been modified in order to make them efficient and suitable for implementation.

Furthermore, this paper proposes a new approach for determining the battery life cycle in function of depth of discharge.

2. STANDARDIZATION WORK

In the literature, one can find a series of test procedures in order to investigate the performances of batteries. Classical standards such as the IEC 60254 for lead batteries [4] typically use constant current discharge cycles. Dynamic discharge profiles have been introduced in the IEC 61982-2 [30], typical for nickel-based batteries in battery electric vehicles. However, existing battery standards are not appropriate for the emerging technologies such as lithium-ion batteries.

New work is being performed on lithium-ion batteries, both by ISO TC22 SC21 and IEC TC21/SC21A/TC69, with several documents being published (ISO 12405-1/2, IEC 62660-1/2, [6,7,8,9]) describing specific test procedures to determine their applicability in HEV and BEV applications. The issues of these standards cover several topics: performance, reliability and safety. This paper handles the performance issues and life cycle test methods. Other issues (no-load capacity, capacity loss at storage, cranking power at low & high temperature) as well as safety issues fall outside of the scope of this paper.

The unusual fact that both IEC and ISO are working on similar documents is a typical illustration of the awkward split of the electrically propelled vehicle standardization between vehicle-oriented committees (ISO TC22 SC21) and electrotechnical committees (IEC TC21) which each have their own standardization cultures [10]. In this case an agreement has been reached to have the ISO document consider the battery system from a vehicle point of view, encompassing both the battery cells and their auxiliary equipment such as cell electronics, high voltage circuit and over current shut-off device including electrical interconnections, interfaces for cooling, high voltage, auxiliary low voltage and communication as presented in Figure 3.

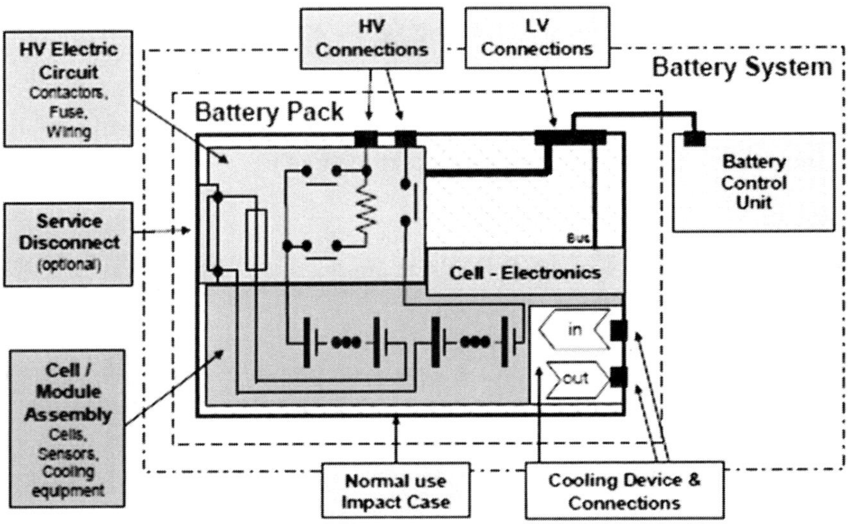

Figure 3. Battery pack system [27].

3. ENERGY AND POWER PERFORMANCES FOR BEV AND HEV APPLICATIONS

3.1. Pre-Conditioning Test

As one can see in Table 1, the specifications of each test procedure according to ISO 12405 and IEC 62660-1 are presented. As defined in Table 1, before starting the real testing sequence, the battery pack (or a cell) should be subjected to some pre-conditioning cycles (five cycles), in order to be sure that an adequate stabilization of the battery pack (or a cell) is reached. The standard considers a battery pack (or a cell) as "preconditioned" when the discharged capacity during two consecutive discharges does not change by a value greater than 3% of the rated capacity. The discharge and charge should to be performed according to the manufacturer's specifications [26,27]. However, the term rated capacity should be interpreted very carefully. In Table 2, we observe that the discharge capacity is about 45 Ah instead of 40 Ah as specified by the manufacturer. Hereby, the rated capacity should be a measured value (e.g., at 2C) when the cell or battery is pre-conditioned.

Table 1. Overview of the defined test in the standard ISO/CD 12405-1/2 and IEC 62660-1.

Test item	Test	Condition	Pack/System ISO/CD 12405-1/2	Extended	Cell IEC 62660-1	Extended
Pre-conditioning	Cycling	Temperature	25 °C		25 °C	
		Charge	standard charge		Standard Charge	
		Discharge	2C		0.2C	$2I_t$
		# Cycles	5		5	
Energy and capacity	CC discharge	Temperature	−18 °C, 0 °C, 25 °C, 40 °C		−20 °C, 0 °C, 25 °C, 45 °C	
		Charge	standard charge	# I_t-rates until max charge I_t-rate (as $1/3I_t$, $1I_t$, $2I_t$, ...)	standard charge	# I_t-rates until max charge I_t-rate (as $1/3I_t$, $1I_t$, $2I_t$, ...)
		Discharge	1C, 10C, 20C, I_{max}	$1/3I_t$, $2I_t$, $5I_t$	1C, 10C, 20C, I_{max}	$1/3I_t$, $2I_t$, $5I_t$
		# cycles	2		2	
Power and resistance	Pulse charge/ discharge	Temperature	−18 °C, 0 °C, 25 °C, 40 °C		−20 °C, 0 °C, 25 °C, 45 °C	40 °C instead of 45 °C
		Discharge	I_{max} dis	$1/3I_t$, $2I_t$, $5I_t$, $10I_t$	0.2C, 1C, 5C, 10C	$1/3I_t$, $2I_t$, I_{max}
		Duration	0.1 s, 2 s, 10 s, 18 s		10 s	
		Charge	$0.75{*}I_{max}$, dis	$1/3I_t$, $2I_t$, $5I_t$, $10I_t$, I_{max}	1/3C, 1C, 5C, 10C	$2I_t$, I_{max}
		Duration	0.1 s, 2 s, 10 s		10s	
		SoC	80%, 65%, 50%, 35%, 20%		50%	80%, 65%, 35%, 20%
Energy efficiency	Pulse charge/ discharge	Temperature	0 °C, 25 °C, 40 °C	−18 °C	−20 °C, 0 °C, 25 °C, 45 °C	40 °C instead of 45 °C
		Discharge	See sequence in Table 4	$1/3I_t$, $1I_t$, $2I_t$, $5I_t$, $10I_t$, I_{max}	1/3C, 1C, 5C, 10C	$2I_t$, I_{max}
		Duration		10s	10s	
		Charge	See sequence in Table 4	$1/3I_t$, $1I_t$, $2I_t$, $5I_t$, $10I_t$, I_{max}	1/3C, 1C, 5C, 10C	$2I_t$, I_{max}
		Duration		10 s	10 s	
		SoC	65%, 50%, 35%	80%, 20%	50%	80%, 65%, 35%, 20%

Table 2. Pre-conditioning test results LFP batteries 40 Ah.

	Cycle 1 (Ah)	Cycle 2 (Ah)	Cycle 3 (Ah)	Cycle 4 (Ah)	Cycle 5 (Ah)
Cell 1	46.83	46.86	46.78	46.69	46.59
Cell 2	45.54	45.50	45.36	45.26	45.15
Cell 3	46.24	46.31	45.25	46.19	46.11

3.2. Energy Capabilities

As mentioned in Table 1, the second performance test is the energy and capacity test. According to this test, different batteries can be compared in terms of discharge capacity at different discharge rates, which is essential in battery electric vehicle applications. This test measures battery (or a cell) capacity in Ah at constant current discharge rates 1C, 10C, 20C andI_{max}. Discharge is terminated on a manufacturer specified discharge voltage limit. However, the numbers of batteries, which are allowed at 10C or 20C are very limited. The discharge current of the most available high energy batteries (or cells) on the market are between 0 and 5C for BEV and up to 15C for HEV applications. In order to take the specifications of these batteries (or cells) in account, the proposed test procedure should be extended to 2C and 5C. It should be noted that the standards specify the current in function of the rated capacity (C) of the battery pack (or a cell). This is a common expression, but a dimensional error taking into account, totally in contradiction with the unit of the current: ampere. In order to solve this, a «reference current» as specified in the standard IEC 61434 can be used [31]:

$$I_t = \frac{C}{1h} \tag{1}$$

The current I_t represents the discharge current in amperes during one hour discharge and C is the measured capacity of a battery pack (or a cell). As presented in Table 1, the test procedure of the capacity and energy test shows different temperatures. In order to compare the performances of a battery pack with the performances of a cell at different temperatures, the temperatures in both cases should be equal and not different as specified in Table 1.

Based on the capacity test, one can calculate the energy density. However, discharging a battery a constant current does not reflect the battery behaviour in BEVs and PHEVs. The discharge performance test presented in Figure 4 is more appropriate. The energy density can be obtained by using the following equation:

$$E = C_{dis} \cdot \frac{V_{aver}}{m} \quad (2)$$

where E is the energy density (Wh/kg) and C_{dis} corresponds to the discharge capacity. V_{aver} represents the average voltage during discharging phase and m is the mass (kg).

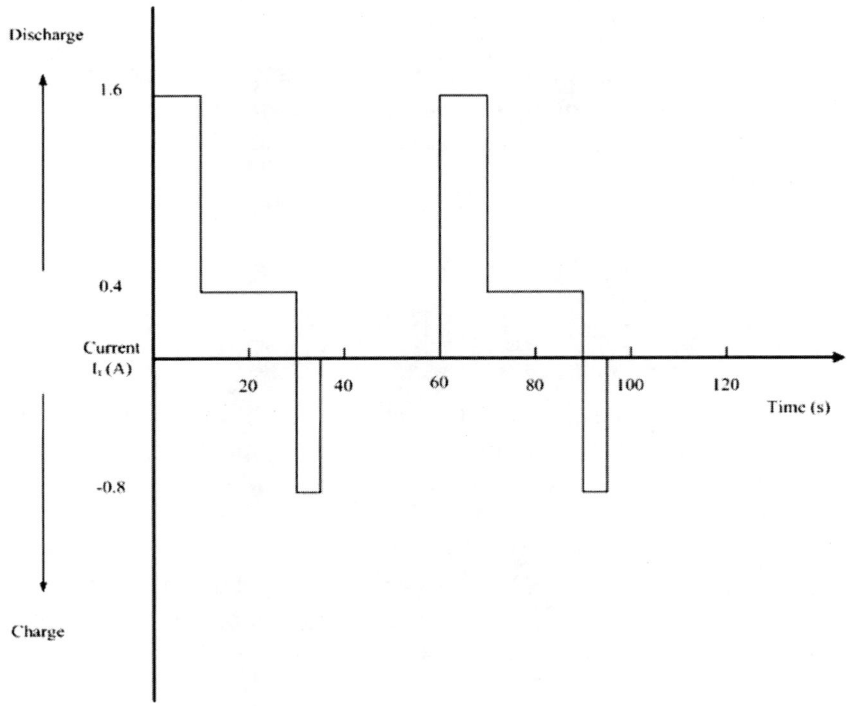

Figure 4. Dynamic Discharge Performance Test [30].

Furthermore, the capacity test does not contain any test sequences regarding charge rate capabilities of batteries. Van den Bossche *et al.* [32] underlined the essential of this aspect due to emerging advance of various charge systems for BEVs and PHEVs. In the framework of this paper, a novel test procedure has been developed (see Figure 5), where the charge and discharge capabilities of a battery cell or pack can be mapped out. In Figure 5, we see that the test presents consecutive charge and discharge phases at different current rates. By this way, the well known Peukert phenomenon can be derived during discharge and the efficiency map at various charge current rates (Figure 6).

The charge efficiency in Figure 6 represents the ratio of stored capacity in Ah at a certain current rate and capacity in Ah at $1I_t$ as presented by Equation (3):

$$Eff_{\text{charge}} = \frac{C_{xIt}}{C_{1It}} \tag{3}$$

The proposed test allows us to estimate the state of charge (SoC) of a battery based on the ampere-hour counting as presented below. The Equation (4) contains all key parameters, which influence the SoC estimation:

$$SoC = 1 - \begin{cases} \dfrac{I_{bat}.T_s}{C.3600}\left(\dfrac{I_{bat}}{I}\right)^{n(LC,t^\circ)-1} & if\ I > 0 \\ -\left[Eff\left(I_{bat}, t^\circ, LC\right).\int I_{bat}.dt\right] & if\ I < 0 \end{cases} \tag{4}$$

with I_{bat} represents the applied battery current in (A), T_s is the step time (s), n, LC, $t°$ are the Peukert constant, life cycle and temperature, respectively. The notation I corresponds to the nominal current, while E_{ff} is the Coulomb charge efficiency.

Figure 7 shows the impact of the Peukert number and the efficiency on the state of charge estimation. It is clear that during real battery operation the battery significantly suffers because of the high current rates.

Further, the standard ISO 12405-1 specify that the test procedure should be performed at least two times. Before starting the real test sequence at different temperatures (see Table 1), an acclimatization step has to be carried out [26,27]. This step will be terminated until a temperature stabilization of ±2 K is reached. From the experience of the author, this step takes around 24 h. Due to the fact that an acclimatization process should happen before starting the energy and capacity test at the temperatures (e.g., −18 °C, 0 °C and 40 °C), this test procedure becomes time consuming.

However, the time duration of this test can be reduced by performing the energy and capacity test separately at the defined temperatures without performing the acclimatization step between the other temperatures.

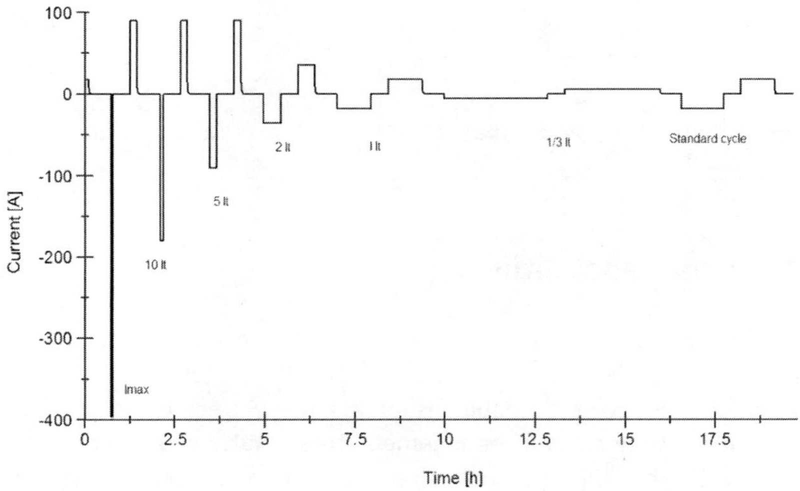

Figure 5. Newly developed energy and capacity test at different current rates.

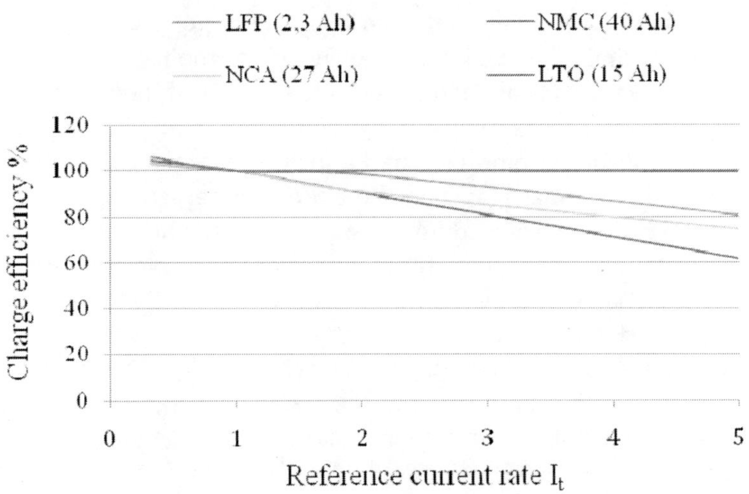

Figure 6. Charge efficiency during CC phase at cell level.

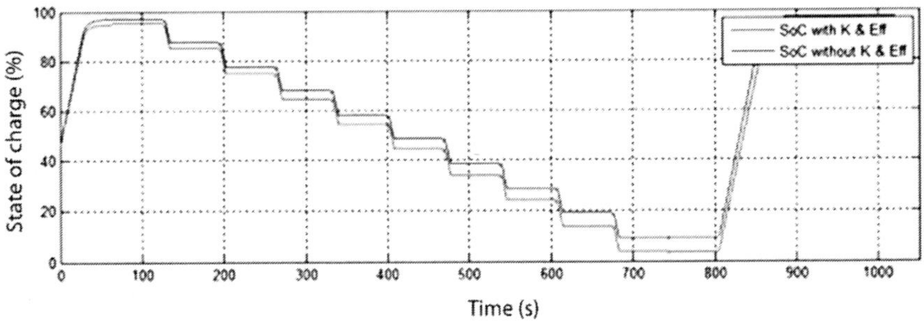

Figure 7. State of charge estimation including Peukert and efficiency.

3.3. Power Capabilities

The third performance test in Table 1 is the hybrid pulse power characterization test, also known as the HPPC test [33]. This test investigates the power capabilities of a battery pack (or a cell) under operating conditions at different temperatures, pulse widths and state of charge (SoC). The HPPC test based on the defined test sequence provides the parameters of several battery models such as Thevenin, FreedomCar and second order FreedomCar, as can be seen in Figure 8.

The objective of this profile is to demonstrate the discharge pulse power (0.1 s, 2 s, 10 s and 18 s) and regenerative charge pulse power (0.1 s, 2 s, and 10 s) capabilities at various SoC and temperatures. The normal test protocol uses the manufacturer's maximum rated pulse discharge current with an upper limitation of 400 A. The current of the regenerative charge pulse shall be kept constant and is calculated at 75% of the discharge pulse current [26,27]. It can be clearly recognized that this test procedure in the present form displays some drawbacks such as acclimatization processes, which are time consuming, as seen in the previous paragraph. Furthermore, the test procedure specifies different pulse widths, which are beyond the allowed performance envelope for most battery packs or cells. Due to the fact that the charge and discharge pulses are not equal, the efficiency of the battery pack or cell cannot be determined since it should be calculated by performing similar tests at equal pulse rates. Finally, the proposals have to be adjusted in the sense that the regenerative charge pulse should not exceed the manufacturer's maximum pulse regenerative voltage. Based on these drawbacks as discussed above and in the previous paragraph, a new test procedure has been developed at the Vrije Universiteit Brussel and Vito Research Institute, which is called the Extended Pulse Power

Characterization Test (Extended HPPC). As one can see in Figure 9, the Extended HPPC test has been modified by multiple equal pulse rates. In the modified form the test can be performed at different equal charge and discharge pulses if permitted by the manufacturer. Due to these modifications, the power capabilities can be calculated at different state of charge values and pulse rates ($1/3I_t$, $1I_t$, $2I_t$, $5I_t$, $10I_t$, $20I_t$ andI_{max}). Further the energy efficiency test as listed in Table 1 can be avoided due to the fact that the pulse rates are equal. The new proposed test procedure is developed only for 10 s pulse because the most batteries or cells are specified at 10 s in analogy with Draft IEC 62660-1. According to the performed modifications, all relevant battery model characteristics can be mapped out.

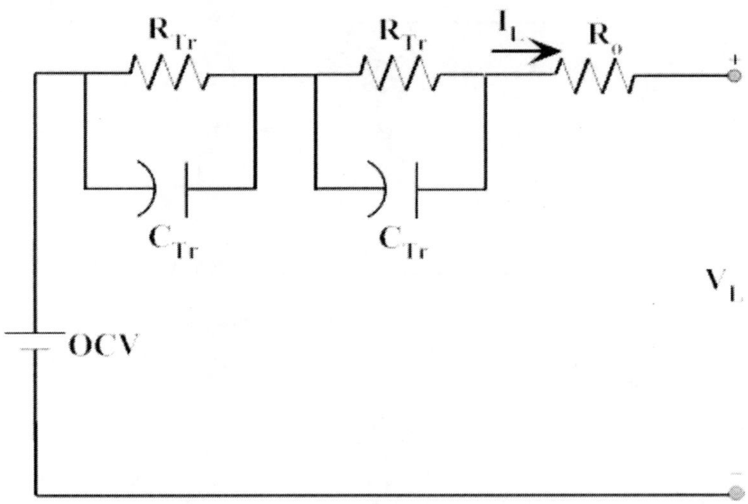

Figure 8. Second order battery model.

It should be noted that the rest period between two consecutive pulses is 10 min instead of 30 min as specified in the ISO 12405-1/2 standards [26,27]. This modification has been down in order to shorten the time schedule of the test procedure. However, this modification does not have an impact on the accuracy of the estimated battery model parameters, as presented in Table 3.

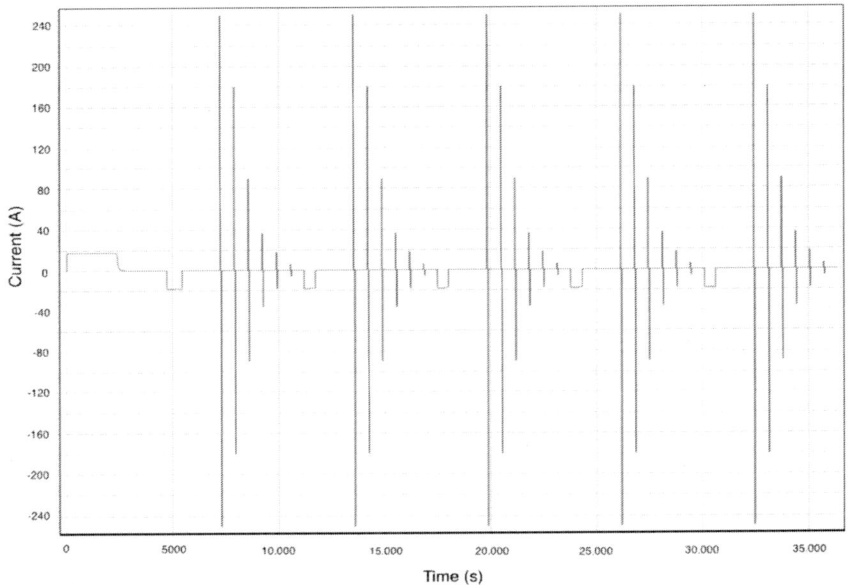

Figure 9. Extended Hybrid Pulse Power Characterization Test.

Table 3. Comparison of the simulation results for both scenarios.

15 min	**Break**					
OCV	OCV'	Ri	Rp1	τ1	Rp2	τ2
3.303	0.017	0.010	0.004	3.109	0.005	44.99
10 min	**Break**					
OCV	OCV'	Ri	Rp1	τ1	Rp2	τ2
3.300	0.018	0.011	0.003	3.102	0.005	44.89

This test is also able to figure out the well-known Butler-Volmer phenomenon. As we can see in Figure 10, the internal resistance decreases the more the current rate increase. However, this result is slightly confusing with the consideration that the energy efficiency decreases at higher current rates. This relationship has been described by Butler-Volmer [34]. General expression of their equation is presented in Equation (5):

$$I=Io\cdot\{e(aFRT\cdot(V-Vo))-e((1-a)FRT\cdot(V-Vo))\} \quad (5)$$

where I_o is the current density, α presents the transfer coefficient, R is the universal gas constant and F is the Faraday constant. While T is the temperature, V and V_o represent the electrode and equilibrium voltages.

As one can see, the equation gives the relationship between the current and the voltage. The voltage can be considered as logarithmically dependent on the current. Hence, the evolution of the battery internal resistance is increasing in function of decreased C-rates. Still, the energy efficiency decreases with increasing current due to quadratic dependence of the loss on the current: RI^2.

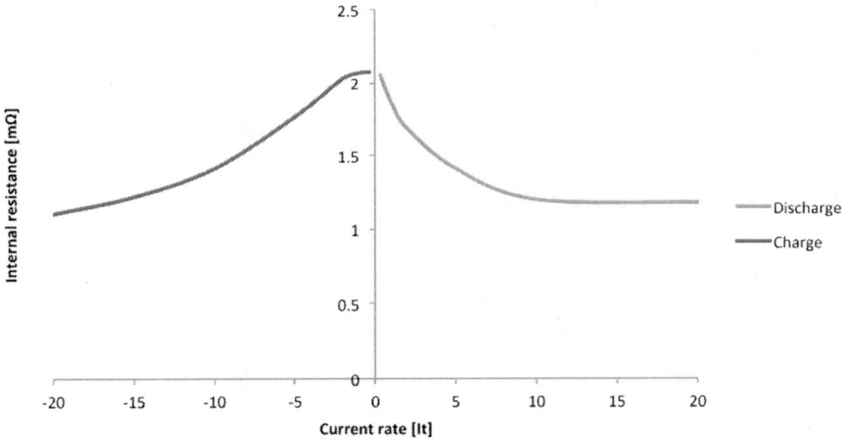

Figure 10. Butler-Volmer phenomenon at lithium-ion cell level.

3.4. Energy Efficiency

The standard IEC 62660-1 at the cell level specify that the energy efficiency test should be carried out by applying equal charge and discharge pulses [28]. However, the ISO 12404-1/2 system-level standards define the energy efficiency test by using the test sequence, as presented by Table 4. In order to investigate the difference between both methods, two tests have been performed at room temperature (RT) with I_{max} equal to $5I_t$ (for a 2.3 Ah LFP battery cell). The experimental results reveal that the energy efficiency according to the IEC 62660-1 is about 89%, while the efficiency based on the other method is 91%. The higher result of the latter one is due to the lower currents that are applied, which have a positive impact on the battery behaviour and efficiency.

Table 4. Test sequence energy efficiency according to ISO/CD 12405-1/2 [28].

Time Increment (s)	Time Cumulative (s)	Current
0	0	0
12	12	$20I_t$ or I_{max}
40	52	0
16	68	$-15I_t$ or I_{max}
40	108	0

4. LIFE CYCLE TESTS

4.1. High Energy

The standard IEC 62660-1 also contains a part about life cycle tests in function of the desired application (battery electric or hybrid electric vehicle). Although the battery electric micro-cycle contains of consecutive charge and discharge pulses, its overall characteristic is charge depleting as can be seen in Figure 11 [28]. The standard IEC 62660-1 specifies that the cell should be subjected to the micro-cycle (Figure 11) until the depth of discharge capacity is 80%, after which the cell will be fully charged. This process of charging and discharging will continue until the cell capacity at $1I_t$ has reached 80% of the initial capacity. The standard specifies that the test should be performed at 45 °C in order to accelerate the ageing mechanism of the test. However, the numbers of commercial lithium-ion batteries that can operate at this charge temperature are very limited [7]. For the most batteries, the charge temperature range is up to 40°C. Cycling the cells (or battery) at higher temperature can lead to damage the battery. However, the results which can be obtained from the mentioned test are not relevant to compare the batteries to each other due to the fact that the life cycle of the cells as specified by the manufacturer are derived at room temperature. The standard purposes to convert the obtained value (45 °C) into a calculated value at room temperature by using the Arrhenius law. However, the Arrhenius law can be used only in the cases when the system is exponential. Due to the fact that the Li-Ion cells are complex rather than exponential, the obtained value according to the standard will not represent a realistic value.

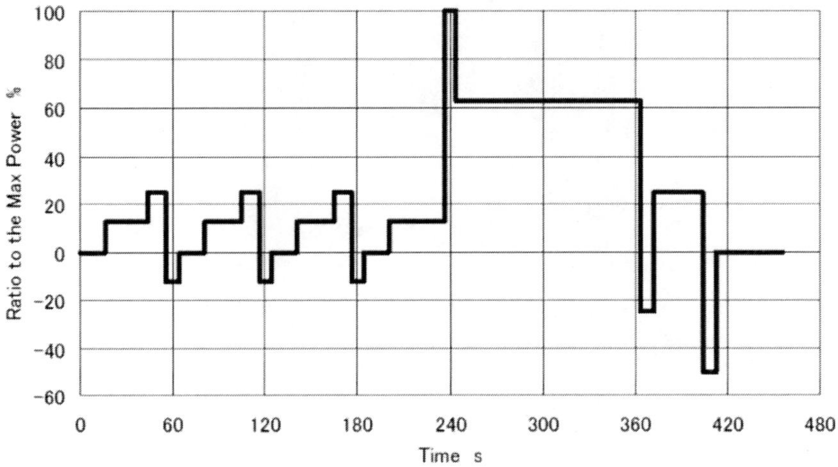

Figure 11. IEC micro-cycle for battery electric vehicle (profileB) [27].

Possible solutions to speed up the ageing mechanisms are the increase of the current rate or the extension of the depth of discharge window. In order to reduce the time duration of the life cycle test, a novel test method has been developed by Punch Powertrain and investigated & optimized at Vrije Universiteit Brussel where the life cycle of a battery can be obtained in function of depth of discharge (DoD). In real applications the depth of discharge has a key impact on the life cycle of a battery (or a cell). In this method, the cells are connected as illustrated in Figure 12. The four connected cells in series will be discharged deeper and undergo higher stress than the other cells. Through this way of connection, one can derive the life cycle in function of depth of discharge DoD. Finally, this test provides also the voltage unbalance in the case of the connected cells in series and current unbalance of the connected stacks in parallel in function of DoD and life cycle. This approach has been examined at laboratory level. The experimental results indicate that the battery cells at each cycle should be balanced. Especially the first four cells in series undergo big differences at high current rates as is presented in Figure 13and Figure 14. Hereby, the discharge capacity during one cycle is limited to 6.44 Ah instead of 9 Ah, which is the nominal capacity. However, the proposed life cycle approach is powerful when a balancing system will be used in real time conditions.

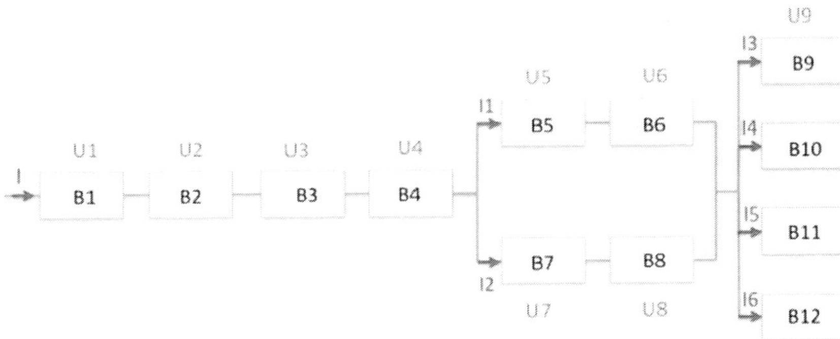

Figure 12. Novel life cycle topology.

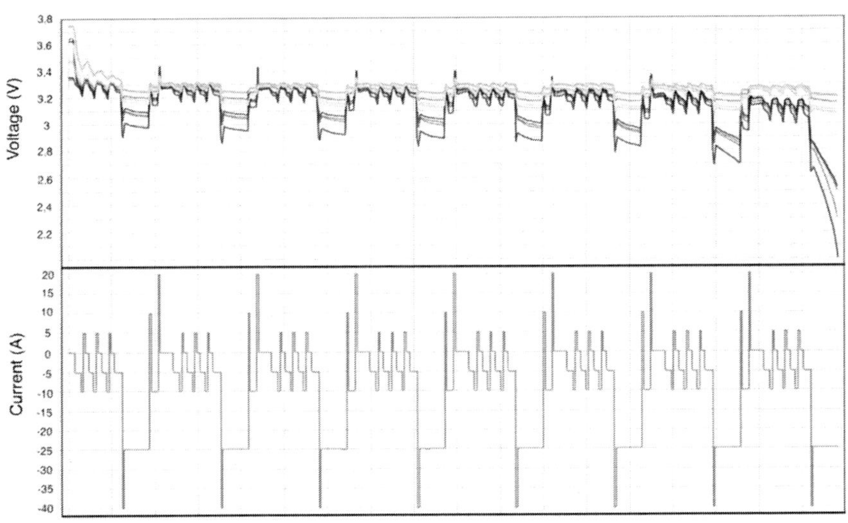

Figure 13. Experimental results of the Punch topology.

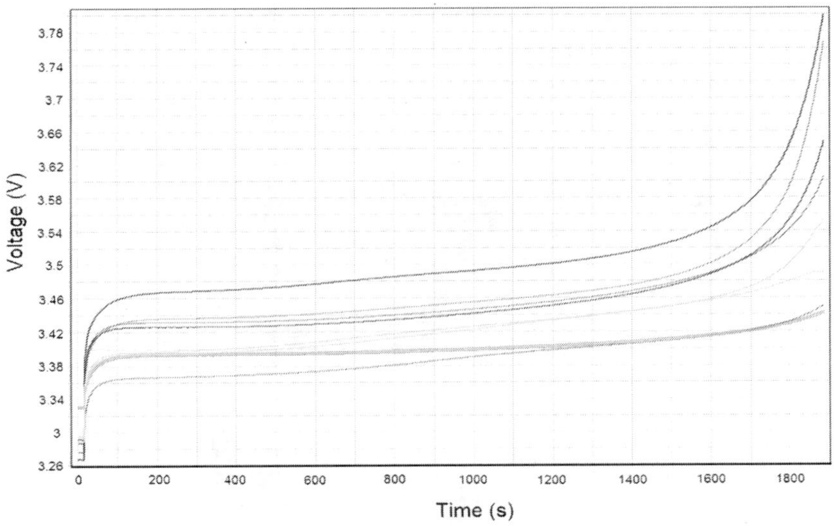

Figure 14. Increasing of variation between the cells during constant current charge without balancing.

4.2. High Power

For hybrid applications, two load profiles have been proposed [28]. They are representing the discharge and charge-rich micro cycles. The discharge cycle will be repeated until 30% DoD has been reached (see Figure 15). Then, the charge sequence will be applied until the upper state of charge limit of 80% SoC is met, as shown in Figure 16. According to the standard, the test should be carried out at 45 °C and the upper discharge and charge current rates are 20 and 12.5I_t, respectively. However, the proposed current rates are too ambitious for most commercial lithium ion batteries. In the framework of a national project in Belgium (Energy Storage project), the life cycle capabilities of seven most promising lithium ion batteries have been investigated at 40 °C. Hereby, it was observed that most battery type as presented in Table 5are not able to recuperate energy at the earlier mentioned current rates and state of charge conditions. As we see in Figure 17, the cell voltage reaches continuously the upper voltage. So the test sequence has been modified whereby the cell where cycled between 70% and 40% SoC and at 10I_t & 7.5I_t discharge and charge, respectively. From the standpoint of the cycling operating window, the proposed values match much better with the typical battery operating conditions in Prius and Honda Civic, where the state of charge varies between 70 and 50% SoC, although these are NiMH batteries and not Li-ion. The tests have been carried out at same temperature as for the high energy (40 °C).

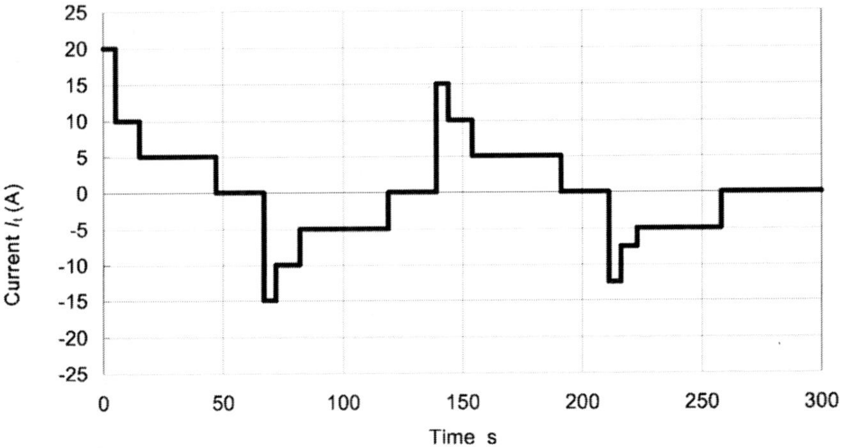

Figure 15. IEC micro-cycle discharge rich profile for battery electric vehicle [28].

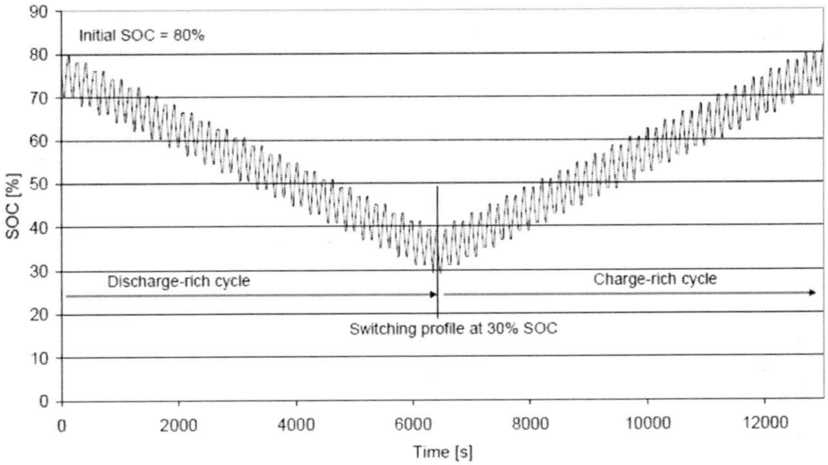

Figure 16. New life cycle method for high power applications [28].

Table 5. Charge and discharge capabilities at 30 and 80% SoC.

Battery	Capacity (Ah)	30% SoC, @ $20I_t$	80% SoC, @ $12.5I_t$
A	2.3	YES	NO
B	7	YES	NO
C	5.7	NO	NO
D	11	NO	NO
E	12	NO	NO
F	10	NO	NO
G	18	YES	YES

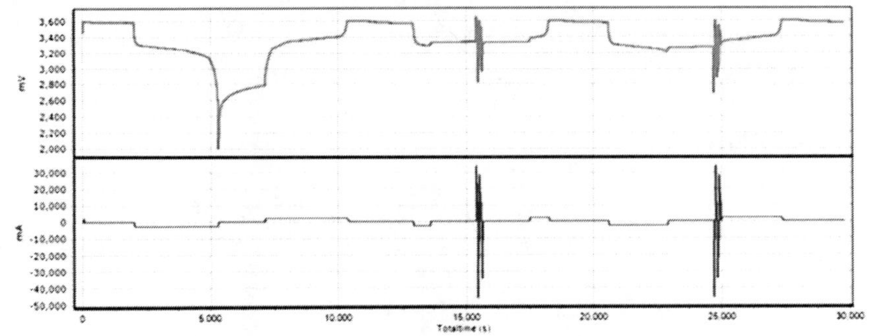

Figure 17. An example of the initial charge and discharge micro cycles.

5. CONCLUSIONS

The study of the standardization process for electricity storage devices allows one to draw interesting conclusions about the general impact of standards, proving on one hand how international standardization does provide a direct benefit to technological and societal development through the deployment of electrically propelled vehicles, and highlighting on the other hand the dynamics of the international standardization world. The proposed work for developing for lithium-ion standards shows in general very interesting test procedures in order to determine the performances of a battery pack/system (or a cell). However, the proposed work should be considered for modifications especially in terms of test efficiency and time duration. During this report, a series of modifications have been proposed in order to make these test sequences more comparable and more efficiently. The improved version of power and resistance test developed at the Vrije Universiteit Brussel makes the proposed test more suited to be used in order to determine the power capabilities of the batteries in a very short time, while integrating the energy efficiency test. The life cycle methodology according to the standard IEC 62660-1 does not have a substantial sense, due to the fact that the Arrhenius law cannot be used to in order to compare the result to the manufacturer value. Therefore a new methodology has been proposed allowing one to derive the life cycle of a battery in function of depth of discharge.

REFERENCES

1. Emadi, A.; Rajashekara, K.; Williamson, S.S.; Lukic, S.M. Topological overview of hybrid electric and fuel cell vehicular power system architectures and configurations. *IEEE Trans. Veh. Technol.* **2005**, *54*, 763–770.

2. Cooper, A.; Furakawa, J.; Lam, L.; Kellaway, M. The ultrabattery—A new battery design for a new beginning in hybrid electric vehicle energy storage. *J. Power Sources* **2009**, *188*, 642–649.

3. Van den Bossche, P.; Vergels, F.; van Mierlo, J.; Matheys, J.; van Autenboer, W. SUBAT: An assessment of sustainable battery technology. *J. Power Sources* **2005**, *162*, 913–919.

4. Axsen, J.; Burke, A.; Kurani, K. *Batteries for Plug-in Hybrid Electric Vehicles (PHEVs): Goals and the State of Technology Circa 2008*; Research Report; Institute of Transportation Studies, University of California, Davis: Davis, CA, USA, 2008.

5. Van Mierlo, J.; Maggetto, G.; van den Bossche, P. Models of energy sources for EV and HEV: Fuel cells, batteries, ultra-capacitors, flywheels and engine-generators. *J. Power Sources* **2004**, *28*, 76–89.

6. Omar, N.; Verbrugge, B.; Mulder, G.; van den Bossche, P.; van Mierlo, J.; Daowd, M.; Dhaens, M.; Pauwels, S. Evaluation of Performance Characterisics of Various Lithium-Ion Batteries for Use in BEV Application. In Proceedings of the IEEE Vehicle Power and Propulsion Conference, Lille, France, 1–3 September, 2010.

7. Omar, N.; Daowd, M.; Verbrugge, B.; Mulder, G.; van den Bossche, P.; van Mierlo, J.; Dhaens, M.; Pauwels, S.; Leemans, F. Assessment of Performance Characteristics of Lithium-Ion Batteries for PHEV Vehicles Applications Based on a Newly Test Methodology. In Proceedings of the 25th World Battery, Hybrid and Fuel Cell Electric Vehicle Symposium, Shenzhen, China, 5–9 November 2010.

8. Burke, A.F. Batteries and ultracapacitors for electric, hybrid, and fuel cell vehicles. *Proc. IEEE* **2007**, *95*, 806–820.

9. Burke, A.; Miller, M. Performance Characteristics of Lithium-ion Batteries of Various Chemistries for Plug-in Hybrid Vehicles. In Proceedings of the 24th World Battery, Hybrid and Fuel Cell Electric Vehicle Symposium, Stavanger, Norway, 13–16 May 2009.

10. Onda, K.; Ohshima, T.; Nakayama, M.; Fukuda, K.; Araki, T. Thermal behaviour of small lithium-ion battery during rapid charge and discharge cycles. *J. Power Sources* **2006**, *158*, 535–542.

11. Yu, H.; Zhang, X.; Jalbout, A.F.; Yan, X.; Pan, X.; Xie, H.; Wang, R. High-rate characteristics of novel anode $LiTi_5O_{12}$/polyacene materials for Li-ion secondary batteries. *J. Electrochim. Acta* **2008**, *152*, 4200–4204.

12. Liaw, B.Y.; Nagasubramanian, G.; Jungst, R.G.; Doughty, D.H. Modeling of lithium ipon cells—A simple equivalent-circuit model approach. *J. Solid State Ion.* **2004**, *175*, 835–839.

13. Wu, M.; Liu, K.H.; Wang, Y.; Wan, C. Heat dissipation design for lithium-ion batteries. *J. Power Sources* **2002**, *109*, 160–166.

14. Sawai, K.; Yamato, R.; Ohzuku, T. Impedance measurements on lithium-ion battery consisting of $Li[Li_{1/3}]O_4$ and $LiCo_{1/2}Ni_{1/2}O_2$. *J. Electrochim. Acta* **2006**, *51*, 1651–1655.

15. Johnson, B.A.; White, R.E. Characterization of commercially available lithium-ion batteries. *J. Power Sources* **1998**, *70*, 48–54.

16. Meekhun, D.; Boitier, V.; Dilhac, J. Study of the Ambient Temperature Effect on the Characteristics and the Lifetime of Nickel-Metal Hybride Secondary Battery. In Proceedings of the Electrical Power & Energy Conference, Montreal, Canada, 22–23 October 2009.

17. Roscher, M.A.; Uwe Sauer, D. Dynamic electric behaviour and open-circuit-voltage modelling of $LiFePO_4$-based lithium ion secondary batteries. *J. Power Sources* **2011**, *196*, 331–336.

18. Wu, H.; Su, C.; Shieh, D.; Yang, M.; Wu, N. Enhanced high-temperature cycle life of $LiFePO_4$-based Li-ion batteries by vinylene carbonate as electrolyte additive. *J. Electrochem. Solid State Lett.* **2006**, *9*, A537–A541.

19. Tikhonov, K.; Koch, V.R. *Li-Ion Battery Electrolytes Designed for a Wide Temperature Range*; Technical Report; Covalent Associates, Inc.: Woburn, MA, USA, 2006.

20. Andre, D.; Meiler, M.; Steiner, K.; Walz, H.; Soczka-Guth, T.; Uwe Sauer, D. Characterization of high-power lithium-ion batteries by electrochemical impedance spectroscopy. II: Modelling. *J. Power Sources* **2010**, *196*, 5334–5341.

21. Lee, S.; Kim, J.; Lee, J.; Cho, B.H. State-of-charge and capacity estimation of lithium-ion battery using a new open-circuit voltage versus state-of-charge. *J. Power Sources* **2008**, *185*, 1367–1373.

22. Dubary, M.; Svoboda, V.; Hwu, R.; Law, B.Y. A roadmap to understand battery performance in electric and hybrid vehicle operation. *J. Power Sources* **2007**, *174*, 366–372.

23. Hu, Y.; Yurkovich, S.; Guezennec, Y.; Yurkovich, B.J. Electro-thermal battery model identification for automotive applications. *J. Power Sources* **2011**, *196*, 449–457.

24. Hu, Y.; Yurkovich, S.; Guezennec, Y.; Yurkovich, B.J. A technique for dynamic battery model identification in automotive applications using linear parameter varying structures. *J. Control Eng. Pract.* **2009**, *17*, 1190–1201.

25. Thele, M.; Bohlen, O.; Uwe Sauer, D.; Karden, E. Development of a voltage-behavior model for NiMH batteries using an impedance-based modelling concept. *J. Power Sources* **2008**, *175*, 635–643.

26. ISO 12405-1: Electrically propelled road vehicles—Test specification for lithium-ion traction battery packs and systems—Part 1: High power applications. 2011. Available online: http://www.iso. org/iso/ catalogue_detail?csnumber=51414(accessed on 18 January 2012).

27. ISO 12405-2: Electrically propelled road vehicles—Test specification for lithium-ion traction battery packs and systems—Part 2: High energy applications. 2011. Available online: http://www. iso.org/iso/ catalogue_detail?csnumber=55854(accessed on 18 January 2012).

28. IEC 62660-1: Electrically propelled road vehicles—Test specification for lithium-ion traction battery packs and systems—Part 1: High power applications. 2011. Available online: http://webstore. iec.ch/Webstore/ webstore.nsf/Artnum_PK/44728 (accessed on 18 January 2012).

29. IEC 62660-2: Secondary batteries for the propulsion of electric road vehicles—Part 2: Reliability and abuse testing for lithium-ion cells. 2011. Available online: http://webstore.iec. ch/Webstore/ webstore. nsf/ Artnum_PK/44727 (accessed on 18 January 2012).

30. IEC 61982-2: Part 2: Dynamic discharge performance test and dynamic endurance test. August 2002. Available online:http://webstore.iec.ch/ webstore/webstore.nsf/Artnum_PK/29145 (accessed on 18 January 2012).

31. IEC 61434: Secondary cells and batteries containing alkaline or other non-acid electrolytes—Guide to designation of current in alkaline secondary cell and battery standards. 1996. Available online: http://webstore.iec.ch/webstore/webstore.nsf/Artnum_PK/21274 (accessed on 18 January 2012).

32. Van den Bossche, P.; Verbrugge, B.; Omar, N.; van Mierlo, J. The Electric Vehicles charged by the Grid: Voltage and Power Levels. In Proceedings of the Plug-In Hybrids and Electric Vehicles (PHEV-09), Montreal, Canada, 28–30 September 2009.

33. *FreedomCAR Battery Test Manual for Power Assist Hybrid Electric Vehicles*; U.S. Department of Energy: Washington, DC, USA, 2003.

34. Kharkats, Y.I.; Sokirko, A.V.; Bark, F.H. Properties of polarization curves for electrochemical cells described by Butler-Volmer kinetics and arbitrary values of the transfer coefficient. *Electrochimica Acta* **1995**, *40*, 247–252

Chapter 9

Battery Management Systems in Electric and Hybrid Vehicles

Yinjiao Xing [1], Eden W. M. Ma [1], Kwok L. Tsui [1,2] and Michael Pecht [1,3,*]

[1]Center for Prognostics and System Health Management (PHMC), City University of Hong Kong, Hong Kong, China

[2]Department of Systems Engineering and Engineering Management, City University of Hong Kong, Hong Kong, China

[3]Center for Advanced Life Cycle Engineering (CALCE), University of Maryland, College Park, MD 20742, USA

ABSTRACT

The battery management system (BMS) is a critical component of electric and hybrid electric vehicles. The purpose of the BMS is to guarantee safe and reliable battery operation. To maintain the safety and reliability of the battery, state monitoring and evaluation, charge control, and cell balancing are functionalities that have been implemented in BMS. As an electrochemical product, a battery acts differently under different operational and environmental conditions. The uncertainty of a battery's performance poses a challenge to the implementation of these functions. This paper addresses concerns for current BMSs. State evaluation of a battery, including state of charge, state of health, and state of life, is a critical task for a BMS. Through reviewing the latest methodologies for the state evaluation of batteries, the future challenges for BMSs are presented and possible solutions are proposed as well.

KEYWORDS

battery management system; lithium-ion battery; state of charge; state of health; state of life

1. INTRODUCTION

From portable electronics to electric vehicles (EVs), batteries are widely used as a main energy source in many applications. Interest in batteries for EVs can be traced back to the mid-19th century when the first EV came into existence [1]. Today, since EVs can reduce gasoline consumption up to 75% [2], EV batteries have gained renewed attention in the vehicle market. Boston Consulting Group [3] has reported that, by 2020, the global market for advanced batteries for electric vehicles is expected to reach US $25 billion, which is three times the size of today's entire lithium-ion battery market for consumer electronics. The U.S. Council for Automotive Research (USCAR) and the U.S. Advanced Battery Consortium (USABC) have set minimum goals for battery characteristics for the long-term commercialization of advanced batteries in EVs and hybrid electric vehicles (HEVs) [4]. To enlarge the market share of EVs and HEVs, safety and reliability are the top concerns of users. However, both of them are subject to not only the battery technology but also the management system for the battery. Therefore, a battery management system (BMS), as the connector between the battery and the vehicle, plays a vital role in improving battery performance and optimizing vehicle operation in a safe and reliable manner. In view of the rapid growth of the EV and HEV market, it is urgent to develop a comprehensive and mature BMS.

Similar to the engine management system in a gasoline car, a gauge meter should be provided by the BMS in EVs and HEVs. BMS indicators should show the state of the safety, usage, performance, and longevity of the battery. Due to volatility, flammability and entropy changes, a lithium-ion battery could ignite if overcharged. This is a serious problem, especially in EV and HEV applications, because an explosion could cause a fatal accident [5]. Moreover, over-discharge causes reduced cell capacity due to irreversible chemical reactions. Therefore, a BMS needs to monitor and control the battery based on the safety circuitry incorporated within the battery packs. Whenever any abnormal conditions, such as over-voltage or overheating, are detected, the BMS should notify the user and execute the preset correction procedure. In addition to these functions, the BMS also monitors the system temperature to provide a better power consumption scheme, and communicates with individual components and operators. In other words, a comprehensive BMS should include the following functions:

- Data acquisition
- Safety protection
- Ability to determine and predict the state of the battery
- Ability to control battery charging and discharging
- Cell balancing
- Thermal management
- Delivery of battery status and authentication to a user interface
- Communication with all battery components
- Prolonged battery life

The current BMSs in both academic research and commercialized products are described in Section 2. Major concerns for current vehicles' BMSs are analyzed in Section 3. Section 4 explains the techniques, algorithms, and methods applied for solving the key concerns with BMSs, while Section 5 proposes and analyzes a BMS with multiple functions. In Section 6, the existing challenges and possible solution schemes are put forward. Finally, conclusions are drawn in Section 7.

2. CURRENT BMSS

Comprehensive and mature BMSs are currently found in portable electronics, such as laptop computers and cellular phones, but they have not been fully deployed in EVs and HEVs. This is because the number of cells in a vehicle's battery is hundreds of times greater than that in portable electronics. Moreover, a vehicle's battery is designed not only to be a long-lasting energy system, but also to be a high power system. In other words, batteries for EVs and HEVs have to provide high voltage and high current. These make BMSs for EVs much more complicated than those for portable electronics.

From a hardware structure perspective, three kinds of topologies have been implemented in BMSs, including centralized, distributed and modular structures [6]. However, the functions of the BMSs in each case are similar. Meissner and Richter [7] proposed a layer structure for battery monitoring, battery state, and battery management. Gold [8] categorized the different functions in a BMS. These concepts can be combined into a generic BMS structure with the basic functions as shown in Figure 1. Various sensors are installed in the battery pack for data acquisition at the monitoring layer. The real-time collected data is used to maintain the system's safety and determine the battery state. The battery state determines the charge time, discharge strategy, cell equalization, and thermal management among the cells, while the state will be passed to the user interface as well.

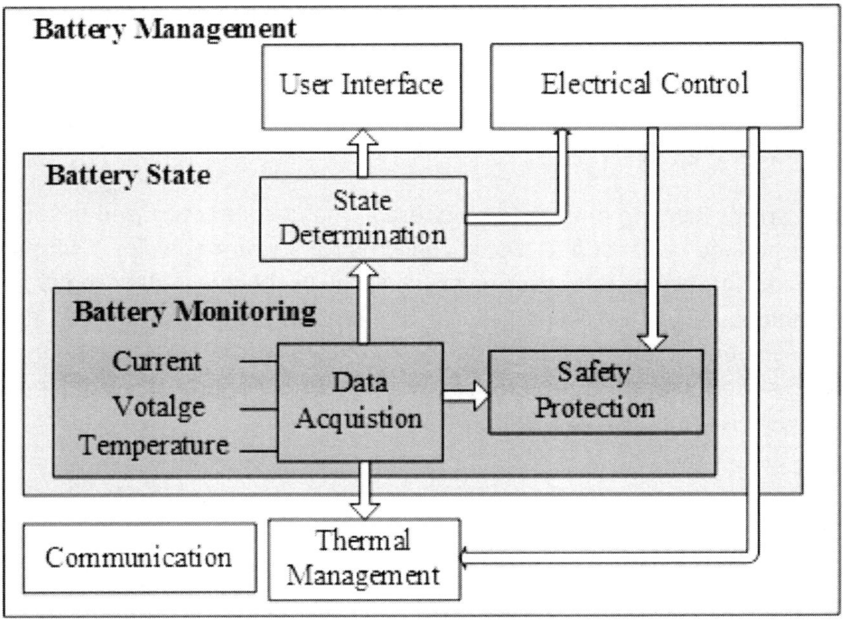

Figure 1. Illustration of a battery management system.

The current commercialized BMSs each perform the basic functions differently. Table 1 shows the main functions of three popular commercial battery management units (BMUs). Drawbacks of the mentioned BMUs include the following:

1) Limited data logging function. The data logging function plays an important role in database establishment, which stores the driving pattern. This profile can help to build up and update the state of charge (SOC) model.

2) Lack of state of health (SOH) and state of life (SOL) estimations. SOH and SOL are used to characterize the current health status and the remaining performance of the battery that will guarantee the reliable operation of the vehicle and scheduled maintenance of the battery replacement.

3) Non-interchangeable among current BMSs. As each BMU has its own cell balancing scheme and communication mechanism, it is impossible to utilize the existing components to form a new BMS.

3. CONCERNS ABOUT VEHICLE BMSS TODAY

With the increasing prices of gasoline and continuing breakthroughs in battery technology, EVs and HEVs were reintroduced in the early 1990s and became mainstream in the 2000s. Because of its promising properties, such as high energy density, long life cycle, and low self-discharge, lithium-ion battery technology has been widely developed and applied in the past decade when the development of BMSs for EVs has been slow and insufficient. This lag has been caused by the following difficulties: (a) battery state evaluation; (b) battery modeling; and (c) cell balancing.

3.1. Battery State Evaluation

Knowledge of the battery state not only helps to determine whether the operational environment is safe and reliable, but also provides information about the charge-discharge operation, which is especially important for cell balancing. Usually, the battery state includes SOC and SOH determination. SOC is similar to the fuel usage indication in gasoline cars, but the battery is inaccessible for measuring and experiences aging, varying environmental conditions, and charge-discharge cycles, which will makes it difficult for a BMS to provide an accurate SOC estimation. According to [12], SOH describes the percentage of battery life remaining. However, there is no consensus on the definition of SOH because it does not correspond to the measurement of a specific physical quality. Although the ratio of the current capacity to the maximum capacity that the battery can hold is usually viewed as a health indicator, more parameters referring to the field performance must be considered during SOH evaluation. The actual formula of the SOH for a specific application is often a trade secret.

SOL is referred to in the literature as the time when the battery must be replaced [12]. It is similar to SOH, but quantifies the remaining time until the battery will be unable to perform. Prediction of battery performance helps the engineer to plan maintenance strategies, and handle disposal and replacement issues.

3.2. Battery Modeling

Establishing a battery model is difficult due to the complicated electrochemical mechanisms of batteries. From the perspective of chemical characteristics, Scrosati and Garche [13] presented voltage-to-capacity profiles of several Li/Li+ materials. For example, $LiFePO_4$ has a long flat trend when charging, while the voltage profiles of $LiMn_3O_4$ and $LiCo_{1/3}Mn_{2/3}O_2$ gradually increase without a flat region. They showed that a generic model for a battery family does not work well for general applications.

Currently, battery modeling for SOC determination is commonly developed from various equivalent circuit (RC network) models, which are distinct for different material characteristics and accuracy requirements [14,15,16]. Cheng [17] and Tremblay [18] adopted the generic battery model that was integrated in MATLAB [19]. However, the generic model is based on the assumption that the internal resistance is constant during charge and discharge cycles. Thus, the accuracy of this model is subject to challenge. While taking into account SOH estimation, the battery degradation model based on capacity fade was simulated and built [20,21,22]. These model parameters were predominantly achieved in terms of the physical characteristics of the specific anode and cathode. However, the external factors, such as environment temperature and discharge current load, will make these stationary models inaccurate in a dynamic environment. As a result, model selection is always focused on in a BMS.

Table 1. Comparison between different BMS products.

	Maxim DS2726 [9]	**TI BQ78PL114 [10]**	**OZ890 [11]**
Measured Cell Parameters	voltage, current	voltage, current, temperature, impedance	voltage, current
Measured Pack Parameters	N/A	N/A	temperature
Safety Protection	▪ over/under voltage ▪ over current ▪ short circuit current	▪ fail-safe operation of pack protection circuits ▪ three power FETs ▪ one secondary safety output fuse	▪ over/under voltage, ▪ over current, short circuit current ▪ over/under temperature
Data Logging	no	on PC-based GUI only	EEPROM
Communication	unknown	PowerLAN, SMBus	Control Area Network (CAN)
SOC/SOH/SOL Estimation	none	SOC	SOC
Dissipative Cell Balancing	charge shunting	N/A	external equalized resistance
Non-dissipative Cell Balancing	N/A	inductive shuttle charge	N/A

3.3. Cell Balancing

In EVs and HEVs, cells are wired in parallel to form a block to satisfy the requirement of high capacity while several blocks (or cells) are connected in series to provide a high voltage [11]. Each cell is distinct due to manufacturing and chemical offset. Thus, the cells in a series have the same current but different voltage. During charging, capacity fade in cells may result in danger if a cell comes to its full charge easily. In other words, it will suffer from overcharging while all the rest of the cells reach their full charge. Similarly, over-discharge may happen on the weakest cell, which will fail before others during the discharging process. When the battery consists of multi-cells in series, it will be subject to a higher failure rate than any single cell due to a series network. To reduce this effect for prolonging the battery life, an effective cell balancing mechanism that would keep the SOC levels of individual cells in a battery pack as close to each other, should be developed.

The mainstream methods of cell balancing can be separated into two kinds: dissipative and non-dissipative. Both of those methods are dedicated to alleviating or even eliminating cell voltage imbalance [23]. However, dissipative equalizers used by resistors facilitate the dissipation of excess energy or current through heat with low efficiency. Non-dissipative equalizers are usually implemented by transformer, inductor or capacitor [6,24]. They are considered more efficient than dissipative equalizers. However, the exchange of charge or energy among cells makes their charge-discharge profile much more complicated than the conventional profiles. These balancing techniques depend on determining the SOC of each individual cell in the battery.

4. METHODOLOGIES FOR BATTERY EVALUATION IN BMSS

Based on the analysis above, it can be seen that the evaluation of battery status is one of the weakest links in BMS and yet it has a large impact on BMS performance. The top concern for EV users is the safety and reliability of the power system in a vehicle. The most important question is whether they will run out of battery power on the road. These issues refer to the estimation and prediction of SOC, SOH, and SOL of the EV battery. Thus, an accurate quantification of the battery status has become one of the most critical tasks for BMSs. In this section, the latest methodologies for battery state estimation and prediction are reviewed.

4.1. State of Charge (SOC)

SOC is critical, but it is not measurable given the current onboard sensing technologies. The ratio of the currently available capacity to the maximum capacity can be expressed as SOC [25], which is calculated by Equation (1):

$$SOC = 1 - \frac{\int i dt}{C_n} \quad (1)$$

where i is the current, and Cn is the maximum capacity that the battery can hold.

SOC reflects the amount of remaining charge that is available to the battery. It is used to determine the driving distance remaining in EVs, while it indicates when the internal combustion engine should be switched on or off in HEVs [26]. Due to the inherent chemical reactions of the battery and different external loads, the maximum capacity of the battery gradually decreases over time. Uncertainty regarding these factors will lead to non-linear, non-stationary battery degradation characteristics.

The most straightforward approach for SOC estimation is Coulomb counting, which characterizes the energy in a battery in Coulombs. This method calculates the capacity of a battery by integrating the current flowing in and out of the battery over time. SOC can be obtained by referring to the calibration point at full charge. However, this reference point (*i.e.*, the initial point) will change due to battery aging and coulombic efficiency. Thus, the reference point must be compensated when operating at practice conditions, and the SOC estimation should be updated under different measured voltage.

Building an accurate table between the discharge capacity and open circuit voltage (OCV) is necessary for achieving SOC. However, this process is time-consuming because it requires a large amount of training data to guarantee the accuracy of the table [27].

The extended Kalman filter (EKF) has been successfully applied for the estimation of SOC in HEV BMSs. Plett [28] developed this method on a first-order RC network in a series of papers. The traditional Kalman filter (KF) is used for linear problems, while EKF linearizes the prediction by using partial derivatives and Taylor series expansion. Windarko [14] also adopted the EKF to estimate SOC based on an electrical model with two series RC networks. However, the EKF cannot deal with a highly non-linear system since the first order Taylor series approximation cannot give enough accuracy in highly non-linear characteristics. Salkind *et al.* [29] utilized Electrochemical Impedance Spectroscopy (EIS) data to estimate SOC with fuzzy logic. Kozlowski *et al.* [12] input impedance parameters from EIS data into a neural network to estimate SOC. The time series model, autoregressive moving average (ARMA), was also implemented in his work. However, impedance measurement, which is one of the common points of the three methods, suffers from high cost, size constraints and measurement sensitivity.

Hansen and Wang [30] applied a support vector machine (SVM) using both classification and regression to estimate the SOC without establishing a battery circuit model. SVM transforms a low-dimensional nonlinear problem into a high-dimensional linear problem. The kernel function in Equation (2) maps the low-dimensional data into high-dimensional data:

$$f(x) = \sum_{i=1}^{N} (\alpha_i - \alpha_i^*) K(x_i, x) \quad (2)$$

where αi and α∗i are selected during the training process to minimize the loss function. The input vector x included both current and voltage parameters. The estimation of SOC was evaluated under steady state and dynamic state. However, parameter tuning is a tedious process for obtaining the optimal SVM model [27].

4.2. State of Health (SOH)

SOH describes the physical condition of a battery, ranging from internal behavior, such as loss of rated capacity, to external behavior, such as severe conditions [12]. Unlike SOC, there is no clear-cut definition of SOH. A general definition of SOH is that it reflects the health condition of a battery and its ability to deliver specified performance compared to a fresh battery [26,31]. The SOH in EV applications is used to characterize the ability to drive a specific distance or range. SOH in HEV applications is a characteristic of the specified power, such as the cranking power from regenerative braking. Scholars and manufacturers use the percentage of nominal capacity as the health threshold of the battery [17]. When the capacity reduces to 80% of the beginning of life capacity after charge-discharge cycling, it is defined as battery failure. However, studies have defined different rules or indicators to quantify the SOH in terms of battery characteristics, test equipment, and different applications.

Pattipati *et al.* [31] combined capacity fade and power fade as health characteristics. Capacity fade indicates the decrease in the driving range with a fully charged battery pack, and power fade indicates the reduced acceleration capability. Both of these features were input into an auto-regressive Support Vector Regression (SVR) model to estimate SOH. Here, the power fade was due to an increase in cell impedance during aging. The total resistance (R=RHF+Rtc) was obtained from EIS data using nonlinear least squares. Figure 2 shows a Randles circuit model of a battery, where RHF and Rtc are the high frequency resistance and the transfer resistance.

Widodo *et al.* [32] proposed a new feature, sample entropy (SampEn), as input data to predict SOH for target vectors of an intelligent system. SVM and its Bayesian version, relevance vector machine (RVM) [33], were used to compare the predictive performance.

$$P = \frac{V^2}{R} \tag{3}$$

$$\text{Power Fade} = 1 - \left(\frac{\text{Power}(k)}{\text{Power}(0)} \right) = 1 - \frac{R(0)}{R(k)} \tag{4}$$

$$\text{Capicity Fade}(\%) = 1 - \left(1 - \frac{\text{Capacity}(k)}{\text{Capacity}(0)} \right) \times 100\% \tag{5}$$

The results also demonstrated that SampEn could serve as an indicator of SOH. SampEn is expressed as Equation (6):

$$\text{SampEn}(m, r, N) = -In \left[\frac{A^m(r)}{B^m(r)} \right] \tag{6}$$

where N is the total number of data points, m is the length of sequences to be compared, r is the tolerance parameters,Bm(r) is the mean value of two similar signal segments that are composed from input vectors with m points, and Am(r)is similar to Bm(r) and will match for m+1 points.

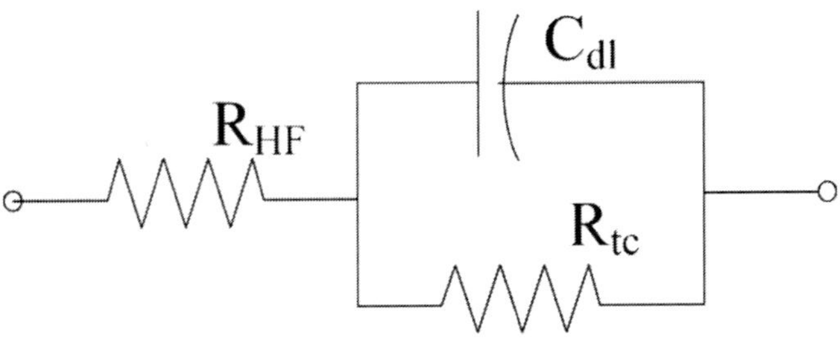

Figure 2. Randles circuit model for a lithium-ion battery.

Chao and Chen [34] designed a state of health estimator for lead-acid batteries. Coup de Fouet voltage [35], internal resistance and transient current, were input into a modified extension matter-element model to develop intelligent SOH evaluation [32]. The mathematical formula is expressed as in Equation (7). Here, health is referred to as the "matter", *R*, which is described by three elements: *N* (name), *C* (characteristic), and *V* (value of characteristic). Through training the data, the weight of each characteristic was quantified and estimated based on the test data. The result was validated and compared with the extension neural network method:

$$R = (N.C.V) = \begin{bmatrix} N\,C_1 V_1 \\ C_2 V_2 \\ C_3 V_3 \end{bmatrix} \tag{7}$$

4.3. State of Life (SOL)

SOL is also known as the remaining useful life (RUL) of a battery. Accurate SOL predictions will facilitate failure prevention and maintenance to prolong the service life of batteries. However, in the past, limited work has been done on SOL prediction. The increasing requirements for battery reliability, especially in military products, have promoted the development of state-of-the-art algorithms.

Pattipati [31] predicted the RUL of a battery using a moving average SVR for different thresholds of capacity fade C(i)and power fade P(i), which are shown in Equations (4) and (5) respectively:

$$RUL(k) = h(\{P(i), C(i)\}_{i=1}^{k}) \tag{8}$$

where k is the kth week. The RUL for an end-of-life criterion is approximately 23% power fade and 30% capacity fade.

RVM can provide the probabilistic interpretation of its outputs. Widodo [32] used RVM regression to predict SOH in capacity fade. The results suggested that RVM produced a more accurate prediction than the SVM model. The likelihood of a data set can be written as Equation (9):

$$p\left(\mathbf{t}|\omega,\sigma^2\right) = \left(2\pi\sigma^2\right)^{N-2} \exp\left\{-\frac{1}{2\sigma^2}\|\mathbf{t}-\Phi\omega\|^2\right\} \tag{9}$$

where Φ is the N×(N+1) design matrix with $\Phi_{nm} = \{1, K(x_i, x_1), K(x_i, x_2), \cdots, K(x_i, x_N)\}^T$; N is the number of input vectors; ω={ω1,ω2,···,ωN} is the regression coefficient vector; ω0 is the bias; and K(x,xi) is the kernel function.

Saha [36] also used an RVM regression model, which was built using internal battery parameters from EIS data. This methodology combined the parameters of an offline model with an online state process. RVM regression was used to fit the related parameters of the exponential models with a probabilistic output. The exponential model is:

$$\tilde{x} = C_x \exp(\lambda_x t) \tag{10}$$

where x˜ is the model's predicted value. In this case, x includes charge transfer resistance, RCT, and electrolyte resistance,RE. The fitted parameters (λRCT, λRE) were input into the online particle filter (PF) process. Based on the combination of PF and RVM, the end of life point can be determined with a narrower probability density function (pdf). However, this approach still relies on the impedance measurements, which are hard to implement due to cost and space constraints.

He [37] developed a degradation model for lithium-ion batteries according to a new exponential model. This finding suggested a new model of the form:

$$Q = a \cdot \exp(b \cdot k) + c \cdot \exp(d \cdot k) \tag{11}$$

where Q is the capacity of the battery; k is the cycle number; a and b are the electrolyte resistance and transfer resistance respectively; and c and d stand for the aging rates. The virtue of this model is that Bayesian Monte Carlo (BMC) updating is performed by regression analysis. The initial values of a, b, c, and d are defined by the weighted sum of the model parameters obtained from the training data. The Dempster-Shafer model is used to get an initial point for BMC updating. Model parameters are updated by BMC to track the degradation trend of the testing battery.

5. A PROPOSED BATTERY MANAGEMENT SYSTEM

The weaknesses of current BMSs are identified through a comprehensive review of the existing approaches. To tackle these weaknesses, we suggest that a comprehensive and mature BMS should contain the components shown in Figure 3 as basic functions.

5.1. Hardware

Safety circuitry has long been used in BMSs. However, since more sensors are used in the proposed BMS, improvements in current safety circuitry designs can be implemented, such as the addition of accurate alarms and controls to prevent overcharge, over-discharge, and overheating.

The sensor system consists of different sensors to monitor and measure battery parameters including cell voltage, battery temperature, and battery current [17]. Some researchers have proposed adopting EIS to monitor internal impedance [12,26]. However, both space constraints and device cost hinder the feasibility of these measurements outside laboratory environments. Thus, current, voltage, and temperature should be measured to improve the capability of state tracking in real life applications.

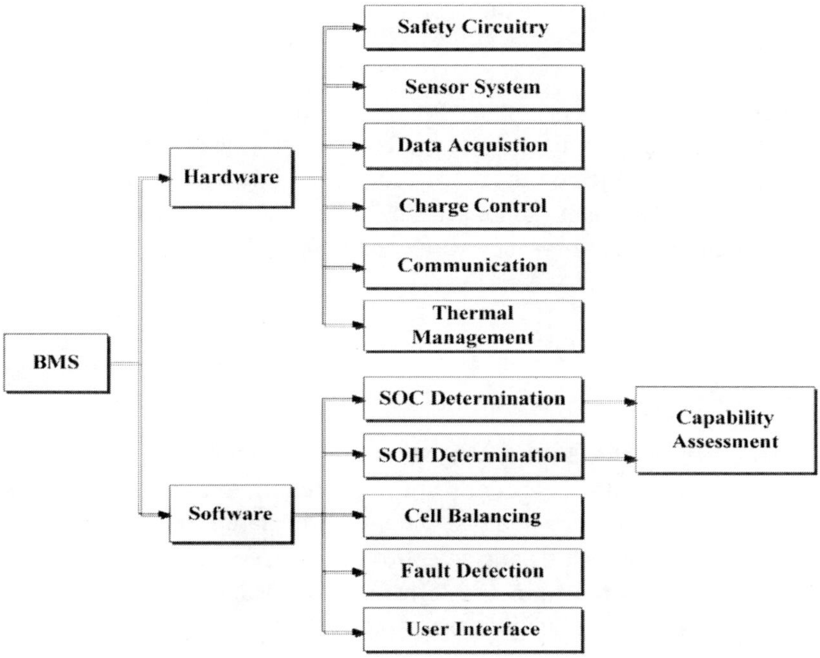

Figure 3. Components of the proposed BMS.

Data acquisition (DAQ) and data storage are critical parts for the software in the BMS to analyze and build a database for system modeling.

Charge control is a subsystem governing the charge-discharge protocol. Batteries are often charged by the constant current/constant voltage method (CC/CV) and will thus need to include a potentiostat and a galvanostat. A variable resistor may be necessary to help balance cells or perform internal resistance measurements. Cell balancing control is still a critical design feature with room for improvement in order to equalize the battery pack and estimate the battery status in an efficient way.

Most subsystems in a BMS are stand-alone modules, and hence, data transfer throughout the BMS is required. Communication through a CAN Bus is a major way to transfer data within the BMS [17,38]. With the development of smart batteries, more data can be collected to communicate with the user and the charger through the microchips incorporated within the battery. In addition, wireless and telecommunication techniques are gradually being incorporated into charging systems that facilitate communication between the battery and the charger.

A module for thermal management is critical because temperature differences have an impact on cell imbalance, reliability and performance. Thus, Pesaran [39] pointed out that it is important to reduce the temperature difference among cells, which must be monitored and operated under proper temperature conditions.

5.2. Software

The software of the BMS is the center of the whole system because it controls all hardware operations and analyses of sensor data for making decisions and state estimations. Switch control, sample rate monitoring in the sensor system, cell balancing control, and even dynamic safety circuit design should be handled by the software of a BMS. Moreover, online data processing and analysis are required for continually updating and controlling battery functions. Reliable and robust automated data analysis is a key factor for success because the analysis determines state estimation and fault detection. This information will be shown to the user through a user-friendly interface with appropriate suggestions. The specific functions of the BMS software are discussed below.

Determination of SOC and SOH will be integrated into a capability assessment, which also presents the life status of the battery and sets the operating limits according to state-of-the-art algorithms, such as fuzzy logic, neural networks, state-space-based models, and so on [17,27,29].

The objective of cell balancing is to maximize battery performance without overcharging or over-discharging. Its nature is to make the SOC levels of cells closer to each other. The controller will control the charge process based on a comprehensive strategy that depends on the SOC of each cell. Thus, the accurate SOC estimation of each cell is basic for improving the balancing.

Most soft faults will be discovered through online data processing. An intelligent data analysis is required in order to provide battery fault warning and indicate out-of-tolerance conditions. Historic data will be recorded and provide the pre-alarm condition before the possible faults.

The user interface should display the essential information of the BMS to the users. The remaining range should be indicated on the dashboard according to the SOC of the battery. Additionally, abnormal alarming and replacement suggestions are needed to inform the users in terms of the estimation and prediction of the battery.

6. CHALLENGES OF BMS AND POSSIBLE SOLUTIONS

A literature review has revealed that BMSs are still in a premature stage [40]. Even if state-of-the-art algorithms and monitoring methods were developed and applied in EVs and HEVs, the reliability of BMSs would still make end users suspicious. Thus, the gap between the laboratory tests and the real requirements should be addressed by future research.

6.1. Challenges

Methodologies to assess battery performance were demonstrated in Section 4 and their characteristics were illustrated as well. Generally, the estimation and prediction methods have unfeasible hardware requirements, such as, impedance measurement, which is costly and not practical in many BMS applications today. Meanwhile, the high computational complexity depends on costly hardware, such as, the central processor. It can be seen that the tradeoff between high performance and feasibility in a BMS is important. Furthermore, most studies are performed in a laboratory environment and are conducted using full charge-discharge cycles. The performance of BMSs under operating conditions, such as vibration from bumpy roads and temperature extremes from snow, rain or summer heat, has rarely been studied. These external loads will be reflected in the battery's available capacity. Thus, it will add un-modeled effects not taken into account in existing algorithms and models. Moreover, with the growth of battery applications, disposal and recycling problems also arise. The problems facing vehicle BMSs are summarized below.

6.1.1. Capacity Estimation under Varying Loads and Environmental Temperatures

Battery degradation models are based on specific materials, environmental conditions, and charge-discharge cycling. Battery status is estimated when discharging at constant current and constant temperature. Ng [25] illustrated the SOC matrix related to the discharge voltage and different discharge rates of lithium-ion batteries (CGR 18650 from Panasonic Co.). However, when it comes to the combined factors, the degradation model based on the single influence factor is subject to query. Figure 4 shows the capacity profile that was tested under two discharge rates and temperatures alternatively.

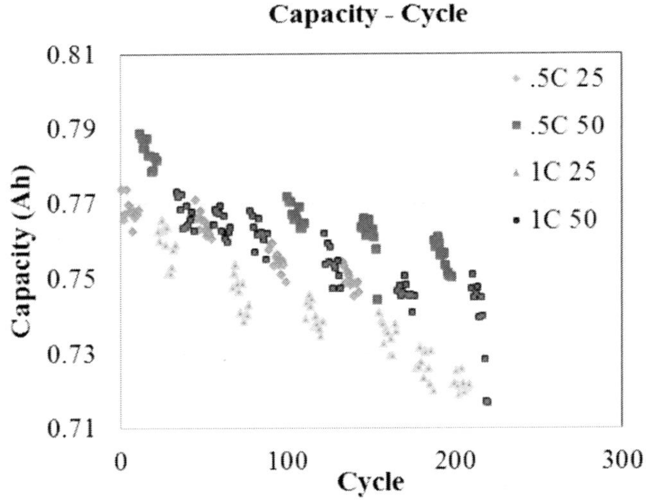

Figure 4. Discharge capacity alternating at the different discharge rates with different temperatures.

The objective of our experiment was to simulate the application under the combined factors. The conditional parameters are listed in Table 2. As Figure 4 shows, the capacity profile moved up with a higher temperature under the same discharge rate, while it went down with a higher discharge rate at the same temperature. Thus, the combined factors increased the complexity of the capacity estimation more than by considering temperature fluctuation or varying discharge current alone.

Table 2. Experiment factors—different discharge rates and temperatures.

Discharge Rate	Temperature
0.5C (350 mA)	25 °C
0.5C (350 mA)	50 °C
1C (700 mA)	25 °C
1C (700 mA)	50 °C

6.1.2. Estimation of Maximum Capacity

The maximum capacity of the battery determines the performance and future life of the battery. The current method for estimation is largely based on the full discharge test. The online capacity is calculated by:

$$\text{Capacity} = \int I dt \tag{11}$$

The longer the integration time, the higher the capacity will be. When the battery is fully discharged at a constant current, the maximum capacity is equal to the result obtained by Equation (11). However, the battery will not be always discharged at a constant discharge current and will not be depleted to the cut-off voltage every time. Figure 5 shows a battery discharged at different depths of discharge. The battery was charged to the full charge, 4.2 V, and stopped discharging at random cut-off voltages. This simulates operation in the real world. Thus, determining how to assess the maximum performance of a battery with partial discharge and variant current loads is a key challenge.

6.1.3. Communication Mechanisms

In practice, the battery needs to communicate with the internal vehicle modules, the charger and external environments through the BMS. The communication between internal modules is relatively mature through a CAN bus, which is standard in vehicles manufactured today [6]. The system management bus (SMBus) was developed to communicate with the charger with the appearance of a smart battery. This battery is able to transmit battery data, such as, the current condition, usage history, and SOC indication through the SMBus. Currently, it is difficult to make the interaction mechanism with the charger uniform due to the different manufacturers and applications. Furthermore, wireless technology is also being developed not only to gather external environment data, such as, humidity and vibration, but also to communicate between the battery and the charger. In such a process, the reliability and interference of data transmission is essential.

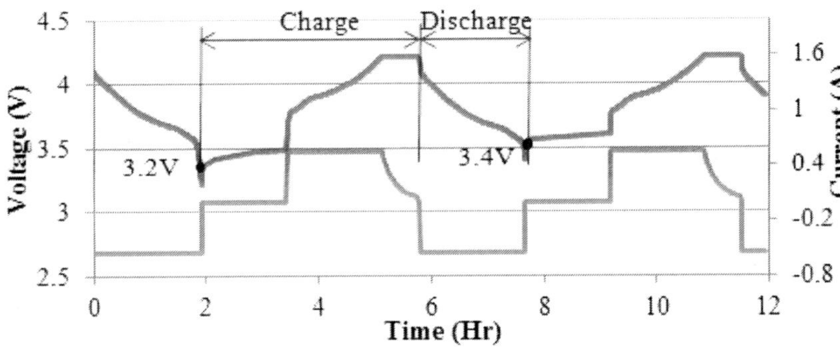

Figure 5. Partial discharge curve.

6.1.4. Assessment of Battery Health

Assessment of battery health is a concern for manufacturers. Since battery technology in EVs is not yet mature, a specific reference or standard needs to be built based on a large database system. In addition to capacity and energy, more weight factors may be combined to evaluate the health status, such as, an increase in cell resistance, a decrease in actual capacity, and the number of charge/discharge cycles.

6.1.5. Battery Disposal and Recycling

As the amount of batteries consumed increases exponentially, disposal has become a significant issue. Tesla has implemented a recycling strategy that reuses or recycles over 60% of a battery's materials [41]. Once pack production volumes increase, this recycling percentage will be raised up to 90%. The battery of Tesla's Roadster consists of 6831 cells that are assembled in 11 sheets. It would be a huge waste of resources if there were no recycling. Unfortunately, from the perspective of most manufacturers, the high recycling costs and the complex disposal process are not worth the effort of implementing an environmentally friendly recycling strategy.

6.2. Possible Solutions for BMSs

Prognostics and health management (PHM) is an enabling strategy consisting of technologies and methodologies for BMSs. By monitoring the sensor signals and processing real-time data from a BMS, the battery status, including SOC, SOH, and SOL, can be estimated and predicted to provide end users with an accurate "gauge meter" in an EV or HEV. Based on the data collected, the BMS determines the corresponding maintenance strategies. Meanwhile, abnormality detection can be used to capture signals to update the predictive results and guarantee the safety and reliability of batteries.

In terms of inaccessible internal reactions of the battery and varying external loads, accurate battery modeling should be established that takes into account imposed factors. Regression technology combined with the state-space models are proposed as a competitive approach for battery degradation modeling. The regression approaches use data training to fit degradation trend curves based on specific battery materials. Once the Markov process utilizes the fitted parameters as its initial information, the empirical degradation characteristics can be combined with the real-time state information to achieve accurate predictive results. Our view is to measure and collect current, voltage, and temperature as the main operational parameters in order to improve feasibility and reduce design costs.

Different driving patterns cause different loads on batteries. The battery power profile can be used to determine the effects of driving behavior on the performance and life of an EV battery. The U.S. Advanced Battery

Consortium (USABC) calculated a battery power-time profile for an Improved Dual Shaft Electric Propulsion (IDSEP) minivan and specified it in the USABC Electric Vehicle Battery Test Procedures [42]. The first step was to map the operational parameters onto the battery loads, such as power or discharge current. A reference point for capacity (*i.e.*, initial fully discharged) was then established to calculate the maximum capacity of the battery.

High costs and the time-consuming nature of these processes are obstacles when it comes to the battery recycling. Due to the large number of surviving cells in a failed battery pack, quick screening methods need to be developed to verify the performance of each cell when recycling a battery pack. It is helpful to classify the unknown batteries in terms of the extent of degradation. Meanwhile, the indication test for a cell with high performance and long life is also important. However, since the battery degradation mechanisms vary in different environmental conditions due to electrochemical characteristics, an accelerating life test is difficult to conduct to estimate battery life at normal conditions. Uno [24] studied the influence of high frequency on the life performance of lithium-ion cells. He showed that high frequency testing had little impact on the calendar life of lithium-ion cells. Thus, the high frequency signal should be developed for screening tests on battery calendar life.

7. CONCLUSIONS

As batteries are the core energy sources in EVs and HEVs, their performance greatly impacts the salability of EVs. Therefore, manufacturers are seeking for breakthroughs in both battery technology and BMSs. Chemical reactions in the battery are subject to operating conditions, and hence, the degradation of a battery may vary in different environments. Developing a comprehensive and mature BMS is critical for manufacturers who would like to increase the market share of their products. The major concerns of BMSs were discussed in this paper. They include battery state evaluation, modeling, and cell balancing, wherein the evaluation methodologies of battery status were viewed as the crucial issue. Thus, related work on the SOC, SOH, and SOL of batteries were reviewed with comparisons. A BMS framework was proposed to deal with the deficiencies of current BMSs in both research and commercial products. Based on previous work, specific challenges facing BMSs and their possible solutions were presented as a solid foundation for future research. Due to varying situations in real-world applications, a standard solution was not wanted. Based on the specific situation, different strategies should be applied to improve and optimize the performance of BMSs in future EVs and HEVs.

ACKNOWLEDGMENTS

The work presented in this paper was fully supported by a grant from the Research Grants Council of the Hong Kong Special Administrative Region, China (CityU8/CRF/09). The authors would like to thank Nicholas Williard and Wei He from CALCE, University of Maryland, for their comments on an earlier version of this article.

REFERENCES

1. Electric Vehicle Wikipedia Page. Available online: http:// en.wikipedia. org /wiki/Electric_vehicle (accessed on 20 July 2011).

2. Claus, D. Materials and processing for lithium-ion batteries. *J. Miner. Metals Mater. Soc.* **2008**, *60*, 43–48.

3. *Batteries for Electric Cars; Challenges, Opportunities, and the Outlook to 2020*; The Boston Consulting Group Inc.: Boston, MA, USA, 2010. Availableonline: http://www.bcg.com/documents/file36615.pdf (accessed on 20 July 2011).

4. *Goals for Advanced Batteries for EVs*; United States Council for Automotive Research LLC: Southfield, MI, USA, 2009. Available online: http://www.uscar.org/commands/files_download.php?files_id=27 (assessed on 1 August 2011).

5. Stuart, T.; Fang, F.; Wang, X.P.; Ashtiani, C.; Pesaran, A. A modular battery management system for HEVs. *Future Car Congress* **2002**.

6. Andrea, D. *Battery Management Systems for Large Lithium Ion Battery Packs*, 1st ed.; Artech House: London, UK, 2010; pp. 22–110.

7. Meissner, E.; Richter, G. Battery monitoring and electrical energy management precondition for future vehicle electric power systems. *J. Power Sources* **2003**, *116*, 79–98.

8. Gold, S. A PSPICE Macromodel for Lithium-Ion Batteries. In Proceedings of IEEE the Twelfth Annual Battery Conference on Applications and Advances, Long Beach, CA, USA, 14–17 January 1997; pp. 215–222.

9. DS2726 Datasheet Rev 3. *MAXIM 5-Cell to 10-Cell Li+ Protector with Cell Balancing*; Maxim Integrated Products, Inc.: Sunnyvale, CA, USA, 2010. Available online: http://datasheets.maxim-ic.com/en/ds/DS2726.pdf (accessed on 1 August 2011).

10. BQ78PL114 Datasheet. *PowerLAN Gateway Battery Management Controller with PowerPump Cell Balancing*; Texas Instruments Inc.: Dallas, TX, USA, 2009. Available online: http://www.ti.com/lit/ds/symlink/bq78pl114.pdf (accessed on 1 August 2011).

11. Intelligent Battery. *The PO_2WER of CO_2OL Battery Technology*; O2Micro International LTD: Grand Cayman, Cayman Islands, UK, 2010. Available online: http://www.o2micro.com/prods/o2m_intbattery_catalog.pdf (accessed on 1 August 2011).

12. Kozlowski, J.D. Electrochemical Cell Prognostics Using Online Impedance Measurements and Model-Based Data Fusion Techniques. In Proceedings of IEEE Aerospace Conference, Big Sky, MT, USA, 8–15 March 2003; Volume 7, pp. 3257–3270.

13. Scrosati, B.; Garche, J. Lithium batteries: Status, prospects and future. *J. Power Sources* **2010**, *195*, 2419–2430.

14. Windarko, N.A.; Choi, J.; Chung, G.-B. SOC Estimation of LiPB Batteries Using Extended Kalman Filter Based on High Accuracy Electrical Model. In Proceedings of Power Electronics and ECCE Asia (ICPE & ECCE), IEEE 8th International Conference, Jeju, Korea, 30 May–3 June 2011; pp. 2015–2022.

15. Yan, J.; Xu, G.; Xu, Y.; Xie, B. Battery State-of-Charge Estimation Based on H_∞ Filter for Hybrid Electric Vehicle. In Proceedings of 10th International Conference on Control, Automation, Robotics and Vision (ICARCV 2008), Hanoi, Vietnam, 17–20 December 2008; pp. 464–469.

16. Zhang, H.; Chow, M.-Y. Comprehensive Dynamic Battery Modeling for PHEV Applications. In Proceedings of IEEE in Power and Energy Society General Meeting, Minneapolis, MN, USA, 25–29 July 2010; pp. 1–6.

17. Cheng, K.W.E.; Divakar, B.P.; Wu, H.J.; Ding, K.; Ho, H.F. Battery-Management System (BMS) and SOC development for electrical vehicles. *IEEE Trans. Veh. Technol.* **2011**, *60*, 76–88.

18. Tremblay, O.; Dessaint, L.A.; Dekkiche, A.I. A Generic Battery Model for the Dynamic Simulation of Hybrid Electric Vehicles. In Proceedings of IEEE Vehicle Power and Propulsion Conference (VPPC 2007), Arlington, TX, USA, 9–12 September 2007; pp. 284–289.

19. *MATLAB/Simulink User's Guide*; The MathWorks Inc.: Natick, MA, USA, 2007.

20. Spotnitz, R. Simulation of capacity fade in lithium-ion batteries. *J. Power Sources* **2003**, *113*, 72–80.

21. Stamps, A.T.; Holland, C.E.; White, R.E.; Gatzke, E.P. Analysis of capacity fade in a lithium ion battery. *J. Power Sources***2005**, *150*, 229–239.

22. Zhang, Q.; White, R.E. Capacity fade analysis of a lithium ion cell. *J. Power Sources* **2008**, *179*, 793–798.

23. Cao, J.; Schofield, N.; Emadi, A. Battery Balancing Methods: A Comprehensive Review. In Proceedings of IEEE Vehicle Power Propulsion Conference (VPPC 2008), Harbin, China, 3–5 September 2008; pp. 1–6.

24. Uno, M.; Tanaka, K. Influence of high-frequency charge–discharge cycling induced by cell voltage equalizers on the life performance of lithium-ion cells. *IEEE Trans. Veh. Technol.* **2011**, *60*, 1505–1515.

25. Ng, K.S.; Moo, C.S.; Chen, Y.P.; Hsieh, Y.C. Enhanced coulomb counting method for estimating state-of-charge and state-of-health of lithium-ion batteries. *Appl. Energy* **2009**, *86*, 1506–1511.

26. Sate of Charge (SOC) Determination. Available online: http://www.mpoweruk.com/soc.htm (accessed on 1 August 2011).

27. Zhang, J.L.; Lee, J. A review on prognostics and health monitoring of Li-ion battery. *J. Power Sources* **2011**, *19*, 6007–6014.

28. Plett, G.L. Extended kalman filtering for battery management systems of LiPB-based HEV battery packs part 2. State and parameter estimation. *J. Power Sources* **2004**, *134*, 277–292.

29. Salkind, A.J.; Fennie, C.; Singh, P. Determination of state-of-charge and state-of-health of batteries by fuzzy logic methodology. *J. Power Sources* **1999**, *80*, 293–300.

30. Hansen, T.; Wang, C.J. Support vector based battery state of charge estimator. *J. Power Sources* **2004**, *141*, 351–358.

31. Pattipati, B.; Pattipati, K.; Christopherson, J.P.; Namburu, S.M.; Prokhorov, D.V.; Qiao, L. Automotive Battery Management System. In Proceedings of IEEE AUTOTESTCON, Salt Lake City, UT, USA, 8–11 September 2008; pp. 581–586.

32. Widodo, A.; Shim, M.C.; Caesarendra, W.; Yang, B.-S. Intelligent prognostics for battery health monitoring based on sample entropy. *Expert Syst. Appl.* **2011**, *38*, 11763–11769.

33. Tipping, M.E. The Relevance Vector Machine. In *Advances in Neural Information Processing Systems*; MIT Press: Cambridge, MA, USA, 2000; Volume 12, pp. 287–289.

34. Chao, K.H.; Chen, J.W. State-of-health estimator based on extension theory with a learning mechanism for lead-acid batteries. *Expert Syst. Appl.* **2011**, *38*, 15183–15193.

35. Pascoe, P.E.; Anbuky, A.H. Standby power system VRLA battery reserve life estimation scheme. *IEEE Trans. Energy Convers.* **2005**, *20*, 887–895.

36. Saha, B.; Goebel, K.; Poll, S.; Christophersen, J. Prognostics methods for battery health monitoring using a bayesian framework. *IEEE Trans. Instrum. Meas.* **2009**, *58*, 291–296.

37. He, W.; Williard, N.; Osterman, M.; Pecht, M. Prognostics of lithium-ion batteries based on dempster–shafer theory and the bayesian monte carlo method. *J. Power Sources* **2011**, *196*, 10314–10321.

38. Jossen, A.; Späth, V.; Doring, H.; Garche, J. Reliable battery operation—a challenge for the battery management system. *J. Power Sources* **1999**, *84*, 283–286.

39. Pesaran, A.A. Battery thermal models for hybrid vehicle simulations. *J. Power Sources* **2002**, *10*, 377–382.

40. Kaiser, R. Optimized battery-management system to improve storage lifetime in renewable energy systems. *J. Power Sources* **2007**, *168*, 58–65.

41. Increasing Energy Density Means Increases Range. Available online: http://www.teslamotors.com/roadster/technology/battery (accessed on 30 July 2011).

42. *Electric Vehicle Battery Test Procedures Manual, Revision 2*; United States Advanced Battery Consortium: Southfield, MI, USA, 1996. Available online: http://www.mdpi.com/1996-1073/4/7/1102/pdf (accessed on 20 July 2011).

Index